CALIFORNIA
DESERT
FLOWERS

The publisher
gratefully acknowledges
the generous contribution to this book
provided by the David B. Gold Foundation
and the General Endowment Fund
of the University of California
Press Associates.

CALIFORNIA DESERT FLOWERS

An Introduction to
Families, Genera, and Species

SIA MORHARDT

EMIL MORHARDT

PHOTOGRAPHS AND ILLUSTRATIONS
BY THE AUTHORS

A PHYLLIS M. FABER BOOK

UNIVERSITY OF CALIFORNIA PRESS

Berkeley   Los Angeles   London

University of California Press
Berkeley and Los Angeles, California

University of California Press, Ltd.
London, England

Produced by Phyllis M. Faber Books
Mill Valley, California

Cover and interior design and typesetting by Beth Hansen-Winter

Library of Congress Cataloging-in-Publication Data

Morhardt, Sia.

    California desert flowers : an introduction to families, genera, and species / Sia Morhardt,
  Emil Morhardt ; photographs and illustrations by the authors.

        p. cm.

    Includes bibliographical references and index.

    ISBN 0-520-24002-2 (cloth: alk. paper) — ISBN 0-520-24003-0 (pbk : alk. paper)

     1. Angiosperm—California—Identification. 2. Desert plants—California—Identification.

QK149.M66 2003
581.7'54'09794--dc21

                                      2003054201

13	12	11	10	09	08	07	06	05	04
10	9	8	7	6	5	4	3	2	1

Printed in Hong Kong through Global Interprint, Santa Rosa, CA

p. ii: *Abronia villosa* (sand verbena), Anza Borrego Desert State Park

p. vi: *Spheralcea ambigua* var. *ambigua* (apricot mallow)

To Camille

Our best work, camping with us in Anza Borrego Desert State Park, 1998

TABLE OF CONTENTS

Young *Ferocactus cylindricus* (barrel cactus) with *Encelia farinosa* (brittlebush)

PREFACE

People have an insatiable appetite for understanding the natural world of which we are a part. Desert flowers have a particular fascination because they seem so fragile and unlikely in their harsh habitat. They can stand out and amaze you or they can be elusive, like the prize in a treasure hunt which finding is half the fun. However you approach the topic of desert flowers, we guarantee that with curiosity and experience your appreciation of every habitat will grow—you will see more, understand more, and be a more interesting companion.

Our photographs will lead you into the world of desert flower families. You will learn the features that tie plant families together and find out how to identify different branches of families—the genera. Imagine that different surnames in your family tree represent genera of plants within a family. Species within a genus are like people with the same surname: they have some things in common but they also have unique features and become easy to recognize once you get to know them. This book is designed to help you recognize common desert flower families, genera, and species, without the need of a large technical vocabulary or a microscope.

Our book is intended to support the explorations of people with an interest in plants or the desert. We hope to have provided an enjoyable guide for making the transition from simple appreciation of flower color to basic knowledge of common plant families and their representatives.

We would like to thank reviewers of our manuscript who were very thorough and helpful, especially Bruce Pavlik, Rodney Myatt, and Phyllis Faber. We particularly want to thank our editor, Doris Kretschmer, for guiding us through the publication process, and her University of California Press associates, including Scott Norton and Jenny Wackman, for attention to detail and helpful suggestions. Anne Canright, our copy editor, made many improvements to the text, and Beth Hansen-Winter, our designer, successfully integrated the numerous photographs and diagrams with our explanations and keys to give this book the eye-appeal we intended. Thanks also to our relatives, friends, and students who looked at early sections of the book and encouraged us along the way.

We gratefully acknowledge support from the Roberts Environmental Center, a research institute at Claremont McKenna College.

Sclerocactus polyancistrus (Mojave fishook cactus)

INTRODUCTION

Everything you need to know about the coverage of our book, about our desert study area, about plant names and parts, and about using keys for identification, plus a key to families, is included in this introduction. If you just want to work on a particular plant family, jump ahead to the appropriate chapter—they are arranged alphabetically by scientific family name. The Table of Contents lists the families we cover.

What You Will Find in this Book

Most Desert Flowers Likely to Attract Your Attention—Twenty-four families of plants with most of the showiest flowers in the desert are included in this book. To keep the book within manageable proportions, we had to leave out some very large and widespread desert families, and we confess to discrimination in favor of beauty and species that are highly characteristic of desert habitats.

Photographs and Descriptions to Hone Your Powers of Observation—There are 351 photographs and diagrams to illustrate families and species described in our book. Chapters are designed to facilitate your recognition of families and their representative genera and species, and to provide you with interesting information.

Simplified Keys for Identifying Flowering Plant Families and Genera—Any flowering plant you find in the Mojave or Sonoran deserts of California is addressed in our keys, usually allowing you to find the family, and often the genus and species. Keys to family are located at the end of this introduction. Keys to genera are located in relevant family chapters.

Icons that Emphasize Important Family Features—Stylized drawings presented at the beginning of each chapter characterize features important in recognizing members of the family, especially desert members. These icons are used as family indicators in the upper corners of the opened book throughout family chapters.

Technical Terms Used Minimally and Illustrated in Locations Where Needed—Only the most basic terms are used in our simplified keys. The terminology needed for using family keys is illustrated or defined in this introduction in the section immediately preceding the family keys. Some families have specialized flower structures with terms useful in the key to genera within that family. These are illustrated or defined in the appropriate family chapters. Use the index to find the page on which a term is defined. The term will appear in bold type on the text of that page or as a label in the figure on that page.

Plant Facts Consistent with the Jepson Manuals—*The Jepson Manual, Higher Plants of California* (Hickman 1996), and its sequel, *The Jepson Desert Manual* (Baldwin et al. 2002), covering the Desert Floristic Province and southern portion of the Great Basin Floristic Province, have provided most of the dimension and distribution data we present. In some cases we have given a slightly modified version based on our own experience with desert populations which may not exhibit the full range of sizes found in a particular species growing under more favorable circumstances. In all cases, we have used plant names as presented in the Jepson manuals. We also have used common names from other books listed in the Annotated References. On rare occasions, when none of these books listed any common name, or when an existing common name was outdated, we slipped in a common name of our own invention.

Our California Desert Study Area

Our book covers flowering plants of the Mojave and Sonoran deserts in California. This includes the entire **Desert Floristic Province** within California, which is a geographic grouping of plants based on the similarity of their origins and representatives. California has two other floristic provinces: the California Floristic Province covering the Sierra Nevada and all western portions of the state, and the Great Basin Floristic Province covering the easternmost portions of the state north of the Mojave Desert. About half of Southern California falls into the Desert Floristic Province, and the province continues east into Nevada and Arizona, and south into Mexico. In California, the **Sonoran Desert** covers almost all of Imperial County, most eastern portions of Riverside County and parts of adjacent counties. Most of Arizona, Baja California and northwestern mainland Mexico also are in the Sonoran Desert. Our little piece of Sonoran Desert is sometimes called the **Colorado Desert** because it falls within the Colorado River Basin where elevations are generally very low and there is a characteristic subset of plants.

In contrast, most of the **Mojave Desert** lies within California. It covers much of San Bernardino and Inyo counties, and spills over into southern Nevada and a bit of Arizona. Most of the Mojave Desert is 600 meters (roughly 2000 feet) or more in elevation, with the highest elevations designated as desert mountains. Our definitions and delineations of these areas agree with The Jepson Desert Manual (Baldwin et al. 2002).

Deserts are arid places, and based on that criterion, most of the Great Basin Floristic Province in California qualifies as a desert as much as the Desert Floristic Province. However, our book does not cover the Great Basin Floristic Province, so when we refer to California's deserts, it is not included. We have chosen to stick with a definition based on floristic provinces because we are talking about plants, and because people think of the Great Basin as a place of sagebrush and piercing winter cold, features that have little in common with most of the Desert Floristic Province.

In general, the Mojave Desert is higher and colder and has less summer rainfall than the Sonoran Desert. Moisture to the Mojave comes mainly from Pacific storms strong enough to cross the Sierra Nevada and other mountains to the west. These storms

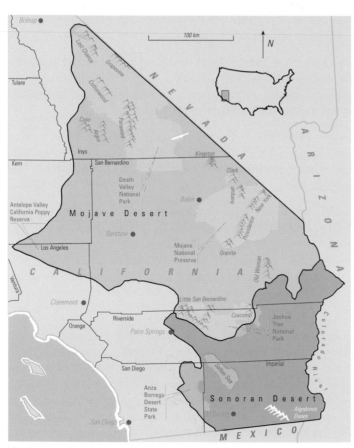

The Desert Floristic Province within California includes both the Mojave and Sonoran deserts.

South end of Death Valley National Park, Amargosa River drainage, 2001

typically are limited to winter months, and may result in snow rather than rain, especially at higher elevations. Even in the southern Mojave Desert occasional freezing temperatures limit the distribution of frost-sensitive species.

The Sonoran Desert receives winter moisture from strong northern Pacific storms, and summer rain from tropical storms coming from the southeast. California's Sonoran Desert sometimes is short-changed on summer moisture because of mountains to the east as well as to the west, but the rainfall pattern is sufficiently reliable to result in a flora somewhat distinct from that of the Mojave Desert.

A rare event in Anza Borrego Desert State Park, Borrego Badlands, 1998

For millions of years plants have responded to a trend toward drier summers in western North America. This long timeframe has allowed evolution to select for mechanisms successful in dealing with seasonal drought. More recently, since the peak of the last ice age, about 22,000 years ago, California has been warming and drying relatively rapidly. Lakes that occupied Death Valley and similar deep basins 10,000 years ago are virtually gone, and plants have shifted too. With less time for evolution to result in physiological changes that keep pace with climate change, existing species have simply shifted in distribution according to their habitat needs.

A few species, such as the California fan palm (*Washingtonia filifera*) have clung to consistently wet habitats such as springs. Several species, such as desert willow (*Chilopsis linearis*) in the largely tropical family Bignoniaceae, are limited to desert washes in which ground water is replenished by seasonal flooding. Other species, once more widespread, are limited to high elevations where temperatures are cool and moisture is reliable. But many species, adapted to drought, have dispersed across vast expanses of desert, finding habitats where temperature, precipitation, and soil conditions meet their needs. Even the salty basins left behind by evaporation of lakes and the shifting sands of dunes support many species with specific adaptations to these habitats.

Creosote bush (*Larrea tridentata*) is a shrub in the Caltrop Family (Zygophyllaceae), that grows throughout California's deserts. It is more drought-resistant than cactus. If you see creosote bush, you are in the desert, almost by definition. If you do not already recognize creosote bush, check out our last chapter, Zygophyllaceae—you probably already know what it looks like and were thinking it was the only thing out there. In the Sonoran Desert, ocotillo (*Fouquieria splendens*) is a defining species. Oddly, it is the only species of its family that occurs in California and, intolerant of severe cold, it is strictly limited to Sonoran Desert. It is a plant you cannot miss—check the chapter on Fouquieriaceae. For the Mojave Desert, it is a lily (Liliaceae), growing like a tree, that is unmistakable evidence you are in the Mojave Desert. You can spot a Joshua tree (*Yucca brevifolia*) from a mile away. There is a chapter for Liliaceae also.

Understanding Plant Names and Classification

Plant Classification—All living things are classified in a Kingdom. Until recently, there were just two: plants and animals. Now there are generally considered to be six Kingdoms, and **Plantae** (plants) is one of them. Organisms such as bacteria, fungi, and algae, have kingdoms of their own and are no longer considered plants. Most people would readily recognize today's Kingdom Plantae organisms as plants, from mosses and ferns to pines and magnolias.

This book deals only with the portion of Kingdom Plantae that includes flowering plants. This is the most recently evolved and advanced division of plants, characterized by flowers and fruits, and generally recognizable as different from ferns and pine trees. Two main subdivisions of flowering plants are recognized: **dicotyledons** (**dicots**) and **monocotyledons** (**monocots**), distinguished mainly by having either two (*di-*) or one (*mono-*) cotyledons—seed leaves—formed in the embryo and sprouting as the seed germinates, before true leaves are fully developed. Cotyledons are a plant's first solar panels, deployed directly from the seed to trap the sun's energy, fueling the process of initial growth essential for survival. Apparently, the distinction of having either one or two cotyledons was made early in the evolution of flowering plants, and it is more consistent than most other anatomical characteristics used to distinguish monocots from dicots.

Dicots and monocots are further divided into subclasses. **Subclasses** are large aggregations of plants with similar features, but there is huge internal variability. Subclasses are made up of **orders**, and orders are divided into **families**. Family features are generally readily recognizable in all members, and family is used as the basis for organization of this and many other books dealing with plant classification. Families are made up of one to many **genera**, and within a genus there are one to many **species**. Species sometimes are further divided into **subspecies** (ssp.) or **varieties** (var.).

Every category (subclass, order, family, etc.) in the classification hierarchy is known as a **taxon**. Every plant belongs in a particular taxonomic hierarchy intended to indicate relationships among groups. However, experts do not necessarily agree to any particular overall scheme of classification. There is a variety of opinions about what organisms constitute which groups and about where different groups should fall in the hierarchy of taxa. Do not expect to find the same presentation in every book you take off the shelf. Plant families are often treated differently in different references, sometimes split apart or sometimes combined from the families we use. Again, we follow the Jepson manuals.

Even if experts agreed on how to interpret evidence, scientific investigation would continue to change our understanding of relationships among organisms. We find previously unknown organisms, both fossil and living, and we find new methods to investigate relationships among organisms. This makes science exciting. Do not let our human desire for categories and rules interfere with your appreciation of the enormous complexity and mystery of the natural world. Enjoy discovery and expect change!

Scientific Names for Plants—Every plant has only one correct scientific name, unique to the species. The name is made up of two Latinized words: the first is capitalized and is the **genus name**, the second is not capitalized and is the **specific epithet**. This **binomial** ("two names") places the plant in a taxonomic hierarchy because each genus belongs to a family, each family to an order, etc.

Genus and species are always written in italics or underlined. Genus names are unique to a single genus but the same epithets may be used with several different genera so they must always appear with the correct genus name (or initial if the full name is obvious from being mentioned earlier). For example, John C. Frémont was an early explorer, leading five expeditions into the Southwest. Several species have been named for him, such as *Lepidium fremontii* (desert alyssum), *Chaenactis fremontii* (Frémont pincushion) and *Populus fremontii* (cottonwood). There is also one genus named for him—*Fremontodendron*, meaning Frémont's "tree" (*dendron*). If another genus were to bear his name it would have to be a variation so that the genus names would remain unique. Often, specific epithets are descriptive of features that distinguish the species from others in the genus, such as *Camissonia cardiophylla*, the *Camissonia* with "heart-shaped" (*cardio-*) "leaves" (*phylla*). The same specific epithet could be used with other genus names so it is not used alone.

Imperfections in this apparently well-organized system arise from historical precedents established prior to international agreement about rules, scientific differences of opinion, and competition for recognition, to name a few sources for confusion. Overall, the first name to be properly described and published has priority, but even the most respected references for a flora may use different names for the same plants, based on different interpretations of which one has priority. In order to clarify the basis for a name, it is often presented not just as a binomial (genus plus specific epithet), but also with the names of the author(s) initiating each portion of the name, usually as an abbreviation(s), not in italics. Thus, *Lepidium fremontii* becomes *Lepidium fremontii* S. Watson. Our book avoids this cumbersome requirement by adopting the same names used in the most recent standard reference for California flora, The Jepson Manual (Hickman 1996).

Pronouncing Scientific Names—Latin is no longer spoken as a regional language, and even if it were, many scientific names are not really Latin, they are merely Latinized, or treated as if they were Latin. For example, plant binomials are sometimes based on a discoverer's name, on a Native American word for the plant, or on Greek descriptive words. Rather than trying to guess origins of names and what that might tell you about pronunciation, say it confidently in Latin, which has no right or wrong pronunciation. However, there are conventions and guidelines, most of which are broken by many respected botanists, so don't stress over it. Helpful generalizations include the following.

1. Imitate people you hear using the words confidently. If you pronounce the word confidently, others are likely to imitate you.
2. Think of the word in terms of many syllables and pronounce all of them as if learning to read. Then say that a little faster, letting the emphasis fall where it seems natural.
3. Where you recognize a root word or especially a proper name, keep its pronunciation intact, not changing emphasis on syllables.
4. Be consistent in your pronunciation of standard endings for subclass, order, and family.

Using Keys for Plant Identification

A key is a written device for matching the features of a plant specimen to its taxonomic group—usually at the family, genus, or species level. Keys generally cover plants in a particular region, such as California or Arizona. Also, keys may be limited to certain groups of plants, such as all flowering plants. Our book includes keys to family and genus for flowering plants of the Mojave and Sonoran deserts of California.

Botanists often can correctly guess the family and genus of an unfamiliar plant species based on patterns of features learned from experience. The botanist will then use a key to species within the appropriate genus, in order to determine species. If you are just learning plant identification, you will need to depend upon a key to families, and within the family, a key to genera, and finally a key to species. Our book has keys and descriptions intended to help you learn to recognize common desert families and genera so that, like the experienced botanist with a new plant, you will eventually need to use keys only to reach the species level.

The keys in this book and many others are based on pairs of statements. Only one statement in each pair is true for any particular plant. Following the path of statements that match the specimen brings you to the name of the taxon that includes the specimen. Our paired statements are numbered. The first member of a pair begins with the numeral followed by a period, and the second member of a pair uses the same numeral, with the same amount of indentation, but it is followed by a prime ('). For the plant in your hand, only one of the statements in the pair is correct. Read both statements, choose, and move on to the choice that follows the correct statement. Continue making choices until you reach the answer.

Our keys were written to be as short as possible so that the number of paired statements is minimal. In part, this is achieved by dividing large groups into smaller ones based on readily distinguished features. Each smaller group then has its own key. Another device for shortening keys is to sometimes allow the user to make a choice from a final list of two or more statements. In this situation, two to five bulleted statements are listed below the last numbered statement. The user can readily see the range of possibilities and choose the single statement true for the plant in hand, and this results in the answer. Done.

Our keys use the most obvious plant features that are reliable and effective. We often provide more than one set of features for making the choice between paired statements. In most cases, these are features seen readily without magnification, and understood readily using normal vocabulary plus the basic terminology presented along with the key. If a particular plant could be interpreted in different ways, or its features are variable with respect to choices in the key, it can usually be found at the end of two or more pathways.

Answers at the end of your trail through family keys are family. If the family you have keyed to is described in our book, its name is printed in bold type. If it is not included, its name is not in bold and you will need a more complete reference to proceed. Genus is given with the family name when a single genus is the only one matching all of the paired statement choices you have made. If more than one genus of a family fits the choices, no genus name is given and you will need to use the genus key in the correct family chapter to make your selection.

If a family has more than one or two genera present in the Mojave or Sonoran deserts of California, our family chapter will include a key to all of them. Any specialized definitions that may be needed to use genus keys are included in the same chapter, usually with diagrams that illustrate specialized structures of the family or distinctions made in the keys. Generic names shown in bold type are those for which we have photographs and descriptions of one or more species. Some genera with large desert representation or spectacular flowers are illustrated by several species.

If a genus key leads you to a genus with a specific epithet, it means that is the only species matching the set of choices you have made in the key. Sometimes only one species of a genus is present in California's deserts. Often, only a genus name is given, in which case it is accompanied by a parenthetical statement of how many species (abbreviated spp for more than one) match the choices made in the key. Other species in the same genus that key out elsewhere are not counted in the number listed. Keys to species are found in the Jepson manuals under the appropriate family and genus.

Essential Flower and Plant Parts and Definitions

Roots, stems, and leaves are a good start at knowing plant parts, but there are a few more essential features to learn before this book will fully serve your classification curiosity. Only the basic and common terms are presented here—sufficient to use the keys to family. In the figures that follow, terms you need to know are printed in black or bolded gray ink. Family chapters cover explanations of any additional terminology needed to appreciate specialized family features or use keys to genera.

Glands are too small to show on our diagrams. They are tiny structures that produce gummy or sticky material, usually on leaves and stems, but often on flower parts as well. Glands may occur at the tips of hairs, directly on plant sur-faces, or embedded slightly under the surface where they may look like dots about the size of a pinhead.

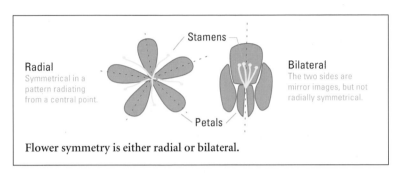

Radial
Symmetrical in a pattern radiating from a central point.

Stamens

Bilateral
The two sides are mirror images, but not radially symmetrical.

Petals

Flower symmetry is either radial or bilateral.

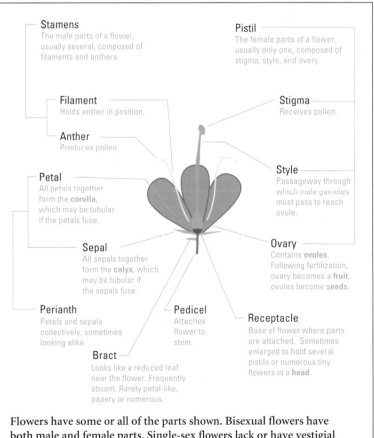

Stamens
The male parts of a flower, usually several, composed of filaments and anthers.

Pistil
The female parts of a flower, usually only one, composed of stigma, style, and ovary.

Filament
Holds anther in position.

Stigma
Receives pollen.

Anther
Produces pollen

Petal
All petals together form the **corolla**, which may be tubular if the petals fuse.

Style
Passageway through which male gametes must pass to reach ovule.

Sepal
All sepals together form the **calyx**, which may be tubular if the sepals fuse.

Ovary
Contains **ovules**. Following fertilization, ovary becomes a **fruit**, ovules become **seeds**.

Perianth
Petals and sepals collectively, sometimes looking alike.

Pedicel
Attaches flower to stem.

Receptacle
Base of flower where parts are attached. Sometimes enlarged to hold several pistils or numerous tiny flowers in a **head**.

Bract
Looks like a reduced leaf near the flower. Frequently absent. Rarely petal-like, papery or numerous.

Flowers have some or all of the parts shown. Bisexual flowers have both male and female parts. Single-sex flowers lack or have vestigial parts from the opposite sex and are called pistillate (female) or staminate (male).

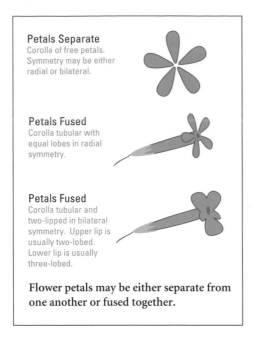

Petals Separate
Corolla of free petals. Symmetry may be either radial or bilateral.

Petals Fused
Corolla tubular with equal lobes in radial symmetry.

Petals Fused
Corolla tubular and two-lipped in bilateral symmetry. Upper lip is usually two-lobed. Lower lip is usually three-lobed.

Flower petals may be either separate from one another or fused together.

Sometimes it is easier to feel the sticky presence of glands than to see them.

Trees are large woody plants with a trunk. The trunk is usually single at ground level with branches above. Trees generally are at least 2–3 m tall and live for many years.

Shrubs are small to medium woody plants with several stems from ground level or multiple branches from near the ground. Shrubs are bushy and grow to only 2–3 m—usually less. Shrubs often live for many years.

Subshrubs are plants with wood only on lower stems. Upper stems are not woody and die back seasonally. Subshrubs generally grow to only about 1 m, but they may live for many years.

Perennials are plants, including trees and shrubs, that live more than two years. However, like the Jepson manuals, we restrict use of the term "perennial" to non-woody plants that live more than two years. Perennials usually are under 1 m tall, but often they live for several years. Generally, they sprout each growing season from a long-lived root system and reduced stem called a **caudex**. Sometimes the caudex is woody, but stems growing from it are not.

Biennials are non-woody plants with a two-year life cycle. In the first year they usually form numerous leaves at ground level but do not flower. The second year, they flower, develop fruits with seeds (if fertilized), and then die. Flowering stalks are generally under 1 m tall.

Annuals are non-woody plants that grow from seed, flower, and produce seeds in a single year or growing season. Seeds are often capable of remaining viable for many years, waiting until conditions are favorable for germination and growth of new plants. Annuals may be tiny to around 1 m tall, with few rare exceptions.

Vines are plants that climb or trail over supporting structures or the ground. They may be either woody or not, and they often have tendrils or other devices for holding on to their support structure.

Monocots are plants having a single cotyledon, or initial leaf-like structure, when they germinate. Older plants usually have flower parts in three's and parallel leaf veins. Wood is unusual and when present tends to be slightly

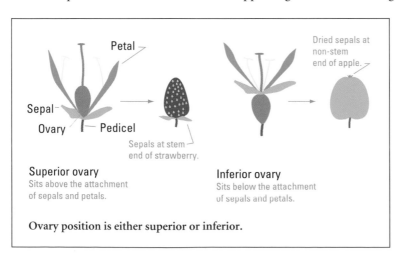

Petal

Sepal
Ovary — Pedicel

Sepals at stem end of strawberry.

Dried sepals at non-stem end of apple.

Superior ovary
Sits above the attachment of sepals and petals.

Inferior ovary
Sits below the attachment of sepals and petals.

Ovary position is either superior or inferior.

spongy. Monocots are one of two large subdivisions of flowering plants that includes palms, grasses, irises, orchids, lilies, and more.

Dicots are plants having two cotyledons. Flower parts often are in four's or five's, but there are many exceptions. Leaves tend to be palmately or pinnately veined, and wood is common but far from universal. When present, wood is typically hard and has seasonal growth rings. Dicots constitute the larger of the two subdivisions of flowering plants and include sunflowers, petunias, beans, oaks, roses, and thousands of others.

Keys to Family

These keys to family cover all 106 families of flowering plants with members in the Mojave and Sonoran deserts of California, outside of cultivated or residential areas. The key to groups guides you to a group key covering a subset of families that share similar obvious features. A few groups with obscure or tiny flowers are not keyed further because their families do not appear in this book. Keys are redundant if there is a likelihood of misinterpreting features. Families included in our book are printed in bold type. Genus is given if it is the only one fitting your choices through the key.

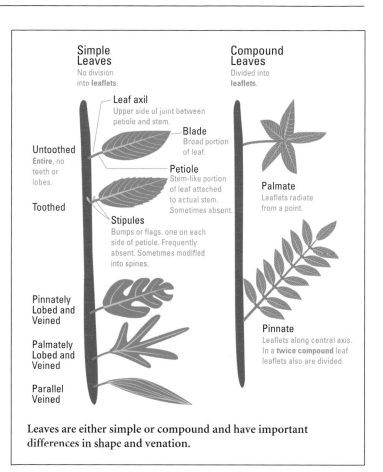

Leaves are either simple or compound and have important differences in shape and venation.

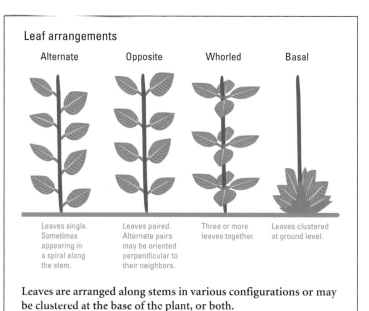

Leaves are arranged along stems in various configurations or may be clustered at the base of the plant, or both.

KEY TO GROUPS

1. Plants aquatic. Leaves all floating or under water (8 families) .... Not Keyed

1' Plants not aquatic. Some growing in wet habitats but with leaves extending above the water.

  2. Plants parasitic. Either living on the stems of other plants and not rooted in the ground, or if rooted in the ground, then with scalelike bracts instead of leaves and no green tissue (5 families) .... Not Keyed

  2' Plants not parasitic, or if so, then appearing normal, with green leaves and roots in the ground.

    3. Leaves with main veins parallel, including huge palm leaves with an overall palmate or pinnate shape. Petals or corolla lobes generally 3 or 6, rarely 4. Petals and sepals often similar or absent. Some flowers single-sex. Monocots and a few dicots (11 families) ....GROUP 1

    3' Leaves generally with pinnate or palmate veins, some with 3 main veins from the petiole but otherwise not parallel veined. Petals or corolla lobes generally 4, 5, or many, some 0, 2, 3 or 6. Dicots.

      4. Trees or large (over 2 m) shrubs with pendant clusters of tiny (5 mm or less) flowers. If male and female flowers grow in separate clusters, only 1 sex need be in pendant clusters.

        5. Leaves compound (*Prosopis*) .... **Fabaceae**

        5' Leaves simple (5 families) .... Not Keyed

      4' Annuals to trees. If trees or large shrubs then not with tiny flowers in hanging clusters.

        6. Annuals to shrubs. Tiny flowers grouped in heads (tight rounded clusters), often appearing to be a single large flower similar to a daisy. Individual flowers without a pedicel or calyx, other parts occasionally missing also. Corollas tubular with 5 (occasionally 0 or 4) lobes, or with 1 elongated petal-like tongue. Ovary inferior, single, 1 ovule. Style forked near the tip, above anthers. Anthers fused around style, or rarely free in male flowers ....**Asteraceae**

        6' Annuals to trees. Generally not with tiny calyx-free flowers clustered in heads. Anthers generally free.

          7. Petals absent (20 families) .... Not Keyed

          7' Petals present. Includes petal-like sepals or bracts with no real petals.

            8. Petals fused into a tube at least at the base. Lobes generally 4 or 5, some 6, some with bilateral symmetry. (Sepals as well as fused petals originate from the receptacle, not from a tubular or dishlike base common to sepals, petals, and stamens.)

              9. Shrubs and trees (18 families) ....GROUP 2

              9' Annuals, perennials, and subshrubs (22 families) ....GROUP 3

            8' Petals separate. Sometimes barely clinging together just at their bases or tips, or fused to a tube of filaments but not to each other. (Some separate petals originate from a tubular or dishlike base shared with sepals and stamens.)

              10. Petals 5 or rarely 7.

                11. Shrubs and trees (18 families) ....GROUP 4

                11' Annuals, perennials, and subshrubs (18 families) ....GROUP 5

10' Petals not 5.

    12. Petals 4 or 8 or more (18 families) ....**GROUP 6**

    12' Petals 1, 2, 3, or 6, or petals and sepals similar and totaling 6 (12 families) ....
....**GROUP 7**

GROUP 1 KEY: Leaves with parallel veins.

1. Trees with a single unbranched trunk like a palm tree. Large, broad leaves at the top ....
.... Arecaceae

1' Generally annuals and perennials. If tree-like, leaves narrow.

  2. Corolla with 4 deep lobes (*Plantago*) .... Plantaginaceae

  2' Corolla absent or with 3 petals or lobes, or appearing as 6 due to similarity of sepals and
   petals.

    3. Flowers showy, petals and sepals white or colorful and generally about 1 cm or more long.

      • Flowers radial, blue. Stamens 3. Leaves grasslike (*Sisyrinchium*) .... Iridaceae

      • Flowers radial, rarely blue. Stamens 6. Leaves rarely grasslike .... **Liliaceae**

      • Flowers bilateral .... Orchidaceae

    3' Flowers not showy. Petals and sepals absent or tiny (generally 3 mm or less).

      4. Petals and sepals absent.

        • Flowers single-sex. Stems generally 3-sided like sedges .... Cyperaceae

        • Flowers single-sex in tight sausage-shaped clusters like cat-tails (*Typha*) ....
.... Typhaceae

        • Flowers generally bisexual, highly modified. Plants like grass .... Poaceae

      4' Petals and sepals together 6.

        • Plants generally intricately branched, not grasslike. Upper leaves generally reduced to
bracts. Stamens 9, rarely 3 or 6 .... Polygonaceae

        • Plants grasslike or reedlike, branched just below flowers at the top of tall round or flat
stems. Stamens 3 or 6 (*Juncus*) .... Juncaceae

        • Plants grasslike, leaves basal. Stem obscure. Stamens 3 or 6 (*Triglochin*) ..... Juncaginaceae

GROUP 2 KEY: Petals fused. Shrubs and trees.

1. Leaves predominantly opposite or whorled.

  2. Flowers single-sex and segregated onto different plants.

    • Leaves generally opposite (*Buddleja*) .... Buddlejaceae

    • Leaves in whorls of 4, less than 1 cm long, tip sharp (*Galium*) .... Rubiaceae

  2' Flowers bisexual.

    3. Ovaries 2 or deeply lobed into almost separate segments with a single ovule per segment.

      • Ovaries 2, stigmas fused .... **Asclepiadaceae**

      • Ovary 2-lobed (*Aloysia*) .... Verbenaceae

• Ovary 4-lobed .... **Lamiaceae**

3' Ovary single and not deeply lobed.

4. Stamens 5, all with anthers .... Caprifoliaceae

4' Stamens 2 or 4 with anthers, some genera with a fifth sterile stamen lacking an anther.

• Stamens 2 .... **Acanthaceae**

• Stamens 4, often with a fifth one that lacks an anther .... **Scrophulariaceae**

1' Leaves predominantly alternate, if somewhat whorled, not small (less than 1 cm).

5. Flowers with bilateral symmetry (*Chilopsis*) .... **Bignoniaceae**

5' Flowers with radial symmetry.

6. Plants with 1 of the following highly distinctive sets of features.

• Bright red flowers along tips of tall, straight, spiny branches (*Fouquieria*) ....

.... **Fouquieriaceae**

• Leaves scalelike, 5 mm or less long, arranged along twigs (*Tamarix*) .... Tamaricaceae

• Leaves pinnately compound .... **Fabaceae**

6' Plants not as in 6.

7. Ovaries 2 or deeply 2-lobed into almost separate segments.

• Ovary deeply 2-lobed. Shrub or subshrub covered with lumpy glands and with a strong odor like turpentine (*Thamnosma*) .... Rutaceae

• Ovaries 2 with styles and stigmas fused. Corolla tube long and erect, surrounding style (*Amsonia*) .... Apocynaceae

• Ovaries 2 with stigmas fused into a thick column. Corolla tube very short, lobes reflexed back toward stem (*Asclepias*) .... **Asclepiadaceae**

7' Ovary single and not deeply lobed.

8. Stamens 9 or 10.

• Corolla with 5 (rarely 4) tiny lobes on a bulbous tube. Stamens 10. Mature branches with smooth rust-red bark (*Arctostaphylos*) .... Ericaceae

• Corolla with 6 deep lobes and small tube. Stamens 9 .... Polygonaceae

8' Stamens 5.

• Corolla generally with 5 lobes, but 4 lobes are common even sharing the same plant with flowers having 5-lobed corollas. Ovary superior (*Lycium*) .... **Solanaceae**

• Apparent corolla with a tube and 5 spreading lobes (the actual sepals) plus 5 tiny upright lobes (the actual separate petals). Ovary inferior (*Ribes*) .... Grossulariaceae

GROUP 3 KEY: Petals fused. Annuals, biennials, perennials, and subshrubs.

1. Ovary inferior or appearing so, or partly inferior.

2. Leaves opposite or whorled.

• Corolla lobes 4. Calyx generally absent, rarely with 4 stiff unequal lobes .... Rubiaceae

• Apparent corolla lobes 3, 5, or more. Calyx appearing to be the corolla and tightly surrounding the ovary. Corolla absent .... **Nyctaginaceae**

2' Leaves predominantly alternate or basal.

3. Vines with tendrils .... **Cucurbitaceae**

3' Not vines, no tendrils.

• Corolla 10–20 mm, radial, mainly tubular, greenish (*Eucnide*) .... **Loasaceae**

• Corolla 2.5–11 mm, radial. Petals mostly separate but fused at the point where the tips broaden and bend outward in a radial pattern, cream-colored (*Petalonyx*) .... **Loasaceae**

• Corolla bilateral, or almost so, fused at the base, generally white, rarely red or yellow. Filaments forming a tube (some free at the base) .... Campanulaceae

1' Ovary superior and appearing so.

4. Ovaries 2 or more or deeply lobed into 2 or 4 segments, rarely with 4 shallow lobes instead of segments.

5. Ovaries 2 or more, if 2 then style and/or stigma joined above.

• Ovaries 2. Corolla tube very short, lobes reflexed back toward the stem (*Asclepias*) .... **Asclepiadaceae**

• Ovaries 2. Corolla tube long and erect, enclosing style (*Amsonia*) .... Apocynaceae

• Ovaries 3–5 .... Crassulaceae

5' Ovary deeply lobed into 2 or 4 segments, rarely with 4 shallow lobes.

6. Leaves opposite.

• Ovary deeply 4-lobed, style attached near base, in the middle of the 4 lobes. Plants generally aromatic, like mint .... **Lamiaceae**

• Ovary deeply 2 or 4-lobed, style attached at ovary tip. Plants often not aromatic .... Verbenaceae

6' Leaves alternate.

• Ovary deeply 2-lobed. Seeds 2 or 4 per lobe (*Menodora*) .... Oleaceae

• Ovary deeply 4-lobed, rarely 2-lobed or with 4 shallow lobes. Seeds 1 (sometimes none) per lobe .... **Boraginaceae**

4' Ovary single and not deeply lobed.

7. Flowers strongly or weakly bilateral.

• Fruit 50–100 mm long and becoming woody with clawlike extensions (*Proboscidea*) .... Martyniaceae

• Fruit 5 mm or less. Stigma lobes 3 (*Loeseliastrum*) .... Polemoniaceae

• Fruit 10 mm or less. Stigma lobes generally 2 or none .... **Scrophulariaceae**

7' Flowers radial.

8. Styles or stigmas 3. Fruits often 3-sided or 3-lobed.

9. Apparent corolla 6-lobed, 3 sepals plus 3 petals usually appearing similar. Leaves generally without stipules, some surrounding stem .... Polygonaceae

9' Corolla or apparent corolla 5-lobed.

• Stamens 5. Leaves without stipules .... Polemoniaceae

• Stamens 3–8. Leaves with stipules that surround the stem forming a membranous ring or sheath (*Polygonum*) .... Polygonaceae

8' Styles or stigmas 1 or 2. Fruits not 3-sided or 3-lobed.

10. Leaves opposite or whorled along stems, some leaves basal.

- Leaves entire. .... Gentianaceae
- Leaves lobed or toothed (*Nemophila*) .... **Hydrophyllaceae**

10' Leaves mainly alternate, some almost opposite, some basal.

  11. Stems climbing or twining like vines, some also upright.

- Leaves palmately lobed or entire, no teeth .... Convolvulaceae
- Leaves pinnately lobed, often toothed (*Pholistoma*) .... **Hydrophyllaceae**

  11' Stems not twining or climbing.

    12. Corolla 4-lobed (*Plantago*) .... Plantaginaceae

    12' Corolla 5-lobed.

- Flowers often lined up along 1 side of a coiled stem tip. Corollas not folded in pleats, not reflexed .... **Hydrophyllaceae**
- Flowers generally in leaf axils or clusters at ends of stems. Corolla generally with pleats where folded in buds. .... **Solanaceae**
- Flowers in clusters at ends of stems. Corolla reflexed with long lobes turned abruptly back toward stem (*Dodecatheon*) .... Primulaceae

GROUP 4 KEY: Petals free, 5 or rarely 7. Shrubs and trees.

1. Flowers bilateral or partially so.

- Fruits spherical with spines. Leaves simple. Magenta petal-like sepals 4 or 5. Petals 3, like tiny flags in a row plus 2 resembling glands (*Krameria*) .... **Krameriaceae**
- Fruits ovoid without spines. Leaves simple (*Polygala*) .... Polygalaceae
- Fruits beanlike. Leaves generally compound (if simple, corollas must be red-purple to blue-purple) .... **Fabaceae**

1' Flowers radial.

  2. Leaves compound.

    3. Leaves twice compound (primary leaflets again subdivided).

- Spiny shrub, not aromatic .... **Fabaceae**
- Unarmed shrub, highly aromatic. Petals white to cream, 5 mm (*Chamaebatiaria*) .... **Rosaceae**
- Unarmed shrub, not aromatic. Petals inconspicuous, stamens huge (*Calliandra*) .... **Fabaceae**

    3' Leaves once compound (leaflets not having leaflets of their own).

      4. Leaflets 3.

- Leaflets 15–30 mm wide, no stipules (*Rhus*) .... Anacardiaceae
- Leaflets 1–4 mm wide, spiny stipules (*Fagonia*) .... Zygophyllaceae

      4' Leaflets 2, or more than 3, pinnately arranged.

- Leaflets 2, joined at the base. Unarmed shrub. Petals yellow, 1 cm (*Larrea*) .... **Zygophyllaceae**
- Leaflets 5–7. Prickly shrub. Petals pink, 2–3 cm (*Rosa*) .... **Rosaceae**
- Leaflets numerous. Short unarmed tree with fat branches (*Bursera*) .... Burseraceae

2' Leaves simple.

    5. Flowers bizarre with narrow petal tips looping back to base of flower (*Ayenia*) .... Sterculiaceae

    5' Flowers without looped petals.

        6. Plants with milky white sap (*Asclepias*) .... **Asclepiadaceae**

        6' Plants with clear or colored sap, not milky white.

           7. Leaves palmately veined and lobed, seasonally deciduous but new leaves in bud at flowering.

             • Woody vine with tendrils like a grape (*Vitis*) .... Vitaceae

             • Tree. Fruits winged in pairs like a maple (*Acer*) .... Aceraceae

             • Shrub, often spiny. Fruits smooth or spiny like currants or gooseberries (*Ribes*) .... Grossulariaceae

             • Shrub, not spiny. Fruits hairy (*Physocarpus*) .... **Rosaceae**

          7' Leaves pinnately veined, some 3-veined from the petiole, but not continuing to spread as in palmate. Some leaves scalelike and extremely sparse.

             8. Stamens 10 or more.

                9. Plants heavily armed with sharply pointed ends on small branches, or with sharp spine-tipped leaves.

                  10. Twigs fat, round, and generally green. Leaves extremely sparse and scale-like (*Castela*) .... Simaroubaceae

                  10' Twigs thin, some angled or turning gray. Leaves oval.

                     • Leaves borne singly along stems. Pistils 1–3 .... Crossosomataceae

                     • Leaves borne in small clusters along stems. Pistil 1 (*Prunus*) .... **Rosaceae**

                9' Plants not armed with sharply pointed twigs or spine-tipped leaves.

                  • Leaves opposite .... Philadelphaceae

                  • Leaves alternate .... **Rosaceae**

             8' Stamens fewer than 10.

                11. Stamens 4 or 5, attached at center of petal base, not between petals .... Rhamnaceae

                11' Stamens 4–9, all or some attached between petal bases (alternate with petals).

                  • Ovary and fruit deeply 2-lobed, style threadlike, stigma tiny. Stamens 8. Plants with turpentine odor (*Thamnosma*) .... Rutaceae

                  • Pistils 1–3, stigma spherical. Stamens 4–10 (*Glossopetalon*) .... Crossosomataceae

                • Pistil 1, stigma with 5 spreading lobes. Stamens 5 (*Mortonia*) .... Celastraceae

GROUP 5 KEY: Petals or petal-like sepals free, 5 or rarely 7. Annuals, perennials, and subshrubs.

1. Flowers distinctly or somewhat bilateral.

    2. Leaves compound .... **Fabaceae**

    2' Leaves simple.

        3. Ovary partly inferior. Leaves basal, broadly 5 to 9-lobed (*Heuchera*) .... Saxifragaceae

3' Ovary fully superior.
- Leaves oval, about 1 cm. Outer sepals 2 and petal-like, cream or pink. Petals 3. Plants thorny (*Polygala*) .... Polygalaceae
- Leaves deeply palmately lobed. Sepals petal-like, blue (rarely white, pink, or red). Petals much smaller than sepals and 4 (*Delphinium*) .... Ranunculaceae
- Leaves broad and heart-shaped. Petals yellow (*Viola*) .... Violaceae

1' Flowers radial.
4. Ovary inferior or partially so.
- Styles 2. Leaves simple with broad lobes and sometimes teeth (*Heuchera*) .... Saxifragaceae
- Styles 2. Leaves intricately dissected or compound .... Apiaceae
- Style 1. Stamens generally many and filaments separate .... **Loasaceae**
- Style 1. Stamens 5 with the upper portions of filaments fused into a tight tube around the style (*Nemacladus*) .... Campanulaceae

4' Ovary, or more than 1 ovary, superior.
5. Ovaries 2 to many.
6. Ovaries 2, stigmas joined .... **Asclepiadaceae**
6' Ovaries 2 to many, stigmas free.
- Leaves simple. Petals less than 3 mm (*Myosurus*) .... Ranunculaceae
- Leaves compound. Petals 4 mm or less .... **Rosaceae**
- Leaves compound to deeply dissected. Petals or petal-like sepals 5 mm or more .... Ranunculaceae

5' Ovary single.
7. Sepals 2 .... Portulacaceae
7' Sepals or sepal lobes more than 2.
8. Leaves consistently alternate, none basal.
- Filaments fused into a column around the style. Fruit 5-lobed or many-segmented. Flowers bisexual .... **Malvaceae**
- Filaments not fused. Fruit 3-lobed. Flowers single-sex (*Ditaxis*) .... Euphorbiaceae
8' Leaves generally opposite. At least some leaves opposite on most plants, others basal, alternate, or whorled.
9. Stamens 6. Petals 4–7.
- Pairs of leaves oriented perpendicular to neighboring pairs (*Frankenia*) .... Frankeniaceae
- Lower leaves opposite, upper alternate (*Lythrum*) .... Lythraceae
9' Stamens 5 or 10 (rarely 9), some with 5 stamens that lack anthers.
- Leaves broad and simple or compound, basal and opposite. Stamens 5 (*Erodium*) .... Geraniaceae
- Leaves generally narrow, opposite in regular pairs. Stamens 10, rarely 9 .... Caryophyllaceae
- Leaves narrow, irregularly arranged. Stamens 5, with or without 5 that lack anthers (*Linum*) .... Linaceae

GROUP 6 KEY: Petals or petal-like sepals free, 4, 8, or more (rarely 0, 5, or 6 on some individuals).

1. Petals or petal-like sepals 8 or more.
- Stems fat, spiny (at least tiny spines), and leafless like a cactus .... **Cactaceae**
- Shrub. Leaves compound, leaflets broad and spiny. Petals 6, plus 9 petal-like sepals, yellow (*Berberis*) .... Berberidaceae
- Perennial. Leaves deeply lobed. Petal-like sepals 5–8, reddish (*Anemone*) .... Ranunculaceae
- Perennial. Leaves entire, strap-shaped. Petals 10–19, white to pink (*Lewisia*) ...... Portulacaceae
- Annual. Leaves toothed. Petals 8, pale yellow (*Mentzelia*) .... **Loasaceae**

1' Petals or petal-like sepals 4 (rarely 0, 5, 6, or 7 on some individuals).
  2. Ovaries 2 to many, or deeply 2-lobed.
  - Ovary deeply 2-lobed. Petals 4. Calyx 4-lobed (*Thamnosma*) .... Rutaceae
  - Ovaries 1–2. Petals 4–5. Sepals free (*Glossopetalon*) .... Crossosomataceae
  - Ovaries 3 or many. Petals 0 or 4. Petal-like sepals 4 or 5 .... Ranunculaceae

  2' Ovary single and not deeply 2-lobed.
    3. Ovary inferior, partially inferior, or appearing inferior.
    - Annuals, fleshy. Sepals 2. Stamens 4–20 (*Portulaca*) .... Portulacaceae
    - Annuals and perennials (rarely subshrubs). Sepals 4. Stamens 8 .... **Onagraceae**
    - Shrub. Sepals 4. Petals 4, white. Stamens many (*Philadelphus*) .... Philadelphaceae
    - Shrub. Petal-like sepals 4, yellow. Petals 0. Stamens many (*Coleogyne*) .... **Rosaceae**

    3' Ovary fully superior and appearing so, or also keyed in 3.
      4. Leaves opposite and numerous (some seasonally leafless).
      - Tree. Leaves palmately compound (*Aesculus*) .... Hippocastanaceae
      - Shrub, upright. Leaves simple, entire (*Coleogyne*) .... **Rosaceae**
      - Subshrub, mat-forming. Wet, alkaline places (*Frankenia*) .... Frankeniaceae
      - Annual. Wet places (*Ammannia*) .... Lythraceae

      4' Leaves alternate or all basal or leaves rare and scalelike.
        5. Shrubs.
          6. Petals or petal-like sepals yellow or magenta.
          - Petals or petal-like sepals magenta. Fruit spherical, spiny (*Krameria*) ...... **Krameriaceae**
          - Petals yellow. Fruit inflated, canoe-shaped (*Isomeris*) .... **Capparaceae**
          - Petals yellow. Fruit not inflated, cylindrical (*Stanleya*) .... **Brassicaceae**
          6' Petals white.
          - Stems leafless and green with dense twigs that are fat, smooth, and tapering to a fearsomely sharp point. Petals 4 (*Koberlinia*) .... Koeberliniaceae
          - Stems sometimes spiny but not also fat, smooth, and green. Petals 4–5. Leaves simple and oval, some deciduous .... Crossosomataceae
          - Stems not spiny. Petals 4. Leaves linear or with linear lobes (*Lepidium*) ...... **Brassicaceae**
        5' Annuals to perennials, rarely subshrubs.

• Sepals 2–3. Stamens 12 to many .... **Papaveraceae**

• Sepals 4. Stamens 6. Ovary on a stalk. Leaves with 3 leaflets .... **Capparaceae**

• Sepals 4. Petals rarely 0. Stamens 6, some 4 or 2. Ovary rarely stalked. Leaves entire, toothed, or pinnately lobed or compound. If compound, not with 3 leaflets .... **Brassicaceae**

GROUP 7 KEY: Petals or petal-like bracts or sepals free, 1, 2, 3, or 6.

1. Petals or petal-like bracts or sepals 1 or 2.

• Petal-like bracts white, 1 per flower, largest (10–30 mm) at the bottom of a tight terminal cluster of flowers with no petals or sepals (*Anemopsis*) .... Saururaceae

• Petals 2, white, 2 mm or less (*Oligomeris*) .... Resedaceae

• Petal or petal-like sepal 1, magenta, bilateral, but in a group of 3 flowers appearing to be radially symmetrical (*Allionia*) .... **Nyctaginaceae**

1' Petals or petal-like sepals 3 or 6.

2. Flowers 3 in a group, appearing to be 3 petals, each with 3 lobes, magenta. Plants trailing on the ground, hairy and glandular (*Allionia*) .... **Nyctaginaceae**

2' Flowers often clustered but not appearing as 1.

3. Shrub. Leaves compound and spiny. Petals 6. Petal-like sepals 9, yellow (*Berberis*) .... Berberidaceae

3' Annuals to shrubs. Leaves entire.

4. Flowers bilateral.

• Stamens 4. Flowers magenta to rose pink. Petals smaller than petal-like sepals and looking like 3 small flags over the center of the flower (*Krameria*) .... **Krameriaceae**

• Stamens 6–8. Flowers cream, or pink and yellow. Petals 3, plus 2 outer sepals forming a pea-like flower (*Polygala*) .... Polygalaceae

4' Flowers radial.

5. Sepals very similar to petals, so sometimes seeming to be absent. Perianth parts 6 (rarely 8 in wet habitats). Flowers often in small clusters.

• Plants growing in water or mud (*Elatine*) .... Elatinaceae

• Plants in drier habitats .... Polygonaceae

5' Sepals generally green and distinct from petals.

6. Sepals 4–6 and free, or tubular with 4–7 lobes. Stamens 6.

• Subshrub, mat-forming. Calyx tubular (*Frankenia*) .... Frankeniaceae

• Perennial, upright. Sepals free (*Lythrum*) .... Lythraceae

6' Sepals 2–3. Stamens 1 to many.

7. Petals 3. Stamens 1–3 (*Calyptridium*) .... Portulacaceae

7' Petals 6, sometimes 4.

• Petals 1–5 mm, yellow. Leaves simple (*Portulaca*) .... Portulacaceae

• Petals 8–40 mm, if smaller then white or leaves finely divided .... **Papaveraceae**

ACANTHACEAE

ACANTHUS FAMILY

Acanthaceae is a large tropical family with only two members that have been successful in our deserts. The single common species is a favorite of hummingbirds. Look for bright red tubular flowers on a large shrub, or look for a hummingbird and follow it.

Acanthos in Greek means "spiny" or "thorny," but California representatives of this family are neither.

Acanthaceae Icon Features

- Club-shaped fruit from a superior ovary
- Seeds (four or two) on launching platforms
- Fruit splitting forcefully to disperse seeds

Justicia californica (chuparosa), near the southern entrance to Joshua Tree National Park

Distinctive Features of Acanthaceae

Acanthaceae species are annuals, perennials, or shrubs with opposite leaves that are simple. Flowers are bilaterally symmetrical with tubular corollas. Technically, Acanthaceae flowers have five corolla lobes, but these are joined so that there seem to be either four or, more commonly, just two.

Acanthaceae members have two (in California) or four stamens with unusual-looking anthers. The ovary is superior, supporting a long single style with one or two stigma lobes. Following fertilization, the ovary develops into a club-shaped fruit designed to toss its ripe seeds as far as possible. This is accomplished by splitting open forcefully, like a clam's shell sitting on its hinge and opening suddenly. There are usually two seeds (sometimes one) in each side of the fruit supported by tiny hooklike braces that direct the seeds' flight away from the bursting fruit.

Similar Families

Some Figwort Family (Scrophulariaceae) members with opposite leaves look enough like Acanthaceae members to be confusing at first. Both have bilaterally symmetrical tubular flowers, superior ovaries, and nearly spherical fruits. However, the potential Figwort Family look-alikes have more than two stamens. The only genus of figworts with just two stamens found in California's deserts is *Veronica*, a small perennial with blue flowers that grows on stream banks in desert mountains. It is impossible to confuse with either species of Acanthaceae in California, which are

subshrubs or shrubs, growing at low elevations in arid habitats.

Two desert genera in the Mint Family (Lamiaceae) have opposite leaves and flowers with just two stamens and bilateral symmetry. However, almost all mints have square stems (4-angled) and are aromatic (i.e., they have an odor like mint or similar), characters not found in Acanthaceae. Also, Mint Family flowers that have structures similar to Acanthaceae flowers are lavender or light blue, distinguishing them from the all-red or white-with-yellow of California's Acanthus Family members.

Family Size and Distribution

There are about 250 genera of Acanthaceae and 3,000 species, distributed throughout the world's tropics. The largest genus is *Justicia*, with about 300 species. Only a few species of Acanthaceae have adapted to arid or temperate conditions. California has only two genera, each with a single species, and both are desert inhabitants. *Justicia* is common at low elevations, and *Carlowrightia* is rare in California but well known in Arizona.

California Desert Genera and Species

Flowers of the two California genera of Acanthaceae are very different. Red corollas (rarely yellow) that have longer tubes than lobes are *Justicia*. White flowers marked with yellow and maroon that have longer lobes than tubes are *Carlowrightia*.

Justicia—This genus was named after a Scottish horticulturalist, James Justice. *Justicia* may be annuals, perennials, or shrubs.

Justicia californica (**beloperone, chuparosa**) is the species "of California" (*californica*). It is a shrub, elaborately branched into a mound, often 1 m or more high. There are few, if any, leaves during most of the year, leaving the gray-green stems to do the work of photosynthesis. The tubular flowers are designed for hummingbirds, which they attract with plenty of nectar. Of course, birds do their part by moving pollen from plant to plant as they feed. The name chuparosa comes from the Spanish verb "to suck," referring to the extraction of nectar. Chuparosa blooms profusely after winter rains, but a few flowers can usually be found almost any time of year. Dried fruits, wide open after discharging their seeds, are often present among fresh flowers. Chuparosa is widespread in sandy washes of the Sonoran Desert below about 800 m.

Examples and Uses

Local Indians, and probably the Spanish too, ate chuparosa flowers as a treat. A few species of Acanthaceae are used for dyes or for medicinal purposes. Several are common ornamentals, such as firespike (*Odontonema*) and yellow shrimp plant (*Pachystachys*). Other than that, they are used mainly by birds and bees.

Opposite page, top to bottom: *Justicia californica* (chuparosa) flowers and dried fruit; unusual yellow coloring of *Justicia californica* (chuparosa); *Justicia californica* (chuparosa) in Anza Borrego Desert State Park, Palm Canyon area

Asclepias erosa (desert milkweed), eastern Mojave near Essex

ASCLEPIADACEAE

MILKWEED FAMILY

Both ornamental and medicinal plants occur in the Milkweed Family. Flower clusters are quite striking, and a beautiful array of seeds, each with a tuft of long silk, can be blown away from dried fruits into the wind like bubbles from a hoop. A few medicines have been derived from milkweed glycosides that affect heart contractions, but these same chemicals are also used in poison arrows and are toxic to grazing livestock. Monarch butterflies, insensitive to the toxins, feed on milkweed and become poisonous to their predators!

Milkweed is a name describing the milky white sap of all members of this family. Asclepiadaceae means this is the family represented by the genus *Asclepias*, named for the Greek god of medicine, Asklepios.

Asclepiadaceae Icon Features

- Central column consisting of fused filaments and anthers enclosing an enlarged style tip and stigma
- Filament appendages often elaborated into hoods and horns
- Petals and sepals generally reflexed

Distinctive Features of Asclepiadaceae

Milkweeds have sticky milky sap, but this feature is not unique to the family. Leaves are usually opposite or in whorls of three or four, but a few species have alternate leaves. Although the family as a whole has various arrangements of flowers, California desert Asclepiadaceae species, with only one exception, have numerous flowers arranged with all their pedicels attached at the same point on the stem, the way the ribs of an umbrella come to the same place on the shaft. Often flower clusters are almost spherical, and although colors are muted, the overall effect is striking. Again, this flower cluster is not unique to Asclepiadaceae, so we must look to individual flowers for distinctive features.

Each flower of Asclepiadaceae is handsomely sculpted and intricately organized. There is a stiff, waxy look to the flowers, and you have to look carefully to distinguish the various parts. Symmetry is radial, with parts in fives, but some of the parts are fused in odd ways and have swollen or hornlike appendages. In the middle of a milkweed flower, a large stigma arises from the fusion of two style tips from two separate superior ovaries. The enlarged stigma and fused style tips are called the **pistil head**. The pistil head is tightly surrounded by five anthers fused into an ornate ring called the **anther head**. Filaments supporting the anthers also are fused, forming a hollow column that encloses the ovaries and portions of the styles. Usually, filaments have large appendages called **hoods** or **horns** depending on their shape. Filament appendages are positioned between petals. Petals often are fused at the base with five lobes spread symmetrically like a five-pointed star behind the elaborate configuration of male and female parts. Sometimes petals are reflexed or folded back along the flower's pedicel, almost as if the elaboration of sexual parts should be allowed a solo performance. Similarly, the five sepals generally are reflexed.

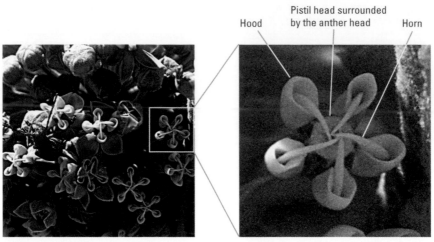

Desert Milkweed (*Asclepias erosa*)

Elaborated sexual parts of Milkweed Family (Asclepiadaceae) flowers in *Asclepius erosa*.

During pollination the elaborate floral structure of milkweeds positions packets of pollen to be picked up by visiting butterflies and sent to neighboring flowers. The pistil head, shrouded by fused anthers, acts as a landing platform for pollinators who come to suck nectar from filament hoods. As the butterfly moves about, its leg slips into a groove between anthers where a saddlebaglike packet of pollen is waiting to attach to the messenger. When the butterfly visits the next flower, the saddlebag is deposited into a receptive stigmatic slit on the side of the pistil head and pollen bursts out of its bags right where it is needed.

Milkweed pods are dried fruits that contain a beautiful set of seeds equipped for flight. The pod splits lengthwise to reveal neatly arranged flat seeds packed like shingles into the stern of the boatlike half pod. Each seed has a tuft of long silky hairs lying flat and extending to fill the prow of the pod. As seeds dry in the sun, their hairs spread like an opening parachute and the seeds take off in the wind. A well-placed puff from a child can launch a flotilla.

Similar Families

A variety of families have at least some desert members with milky sap. These include the Spurge Family (Euphorbiaceae), Sunflower Family (Asteraceae), and Dogbane Family (Apocynaceae), but none has a flower structure resembling the enlarged and sculptured male and female parts of Milkweed Family members. Similarly, a few families such as Lily Family (Liliaceae) and Carrot Family (Apiaceae) have flower clusters similar to those of the milkweeds, but these do not have milky sap or elaborated sexual parts.

Some authorities consider Asclepiadaceae and Apocynaceae to be a single family, in which case it is called Apocynaceae and *Asclepias* is simply a genus in that family. The two families have in common milky sap and paired but separate ovaries with fused styles and/or stigmas, but still, their desert members are easily differentiated. The only Dogbane Family member in California's deserts is *Amsonia tomentosa*, and it has pure white or slightly bluish corollas, each with a long (15 mm) narrow tube and five widely spreading lobes. The calyx lobes stand erect at the base of the corolla tube, and there are no filament appendages. Although the flower has a few odd subtleties, it lacks the waxy, sculpted, and enlarged central features that make Asclepiadaceae species so distinctive.

Family Size and Distribution

There are many opinions as to how many genera are in Asclepiadaceae. The number lies somewhere between 50 and 250, with 2,000 to 3,000 species. Whatever the number, the vast majority of them are found in South Africa and tropical and subtropical South America. *Asclepias* is one of the largest genera, with between 100 and 150 species, and is well represented in temperate climates.

California has five genera of Asclepiadaceae, one of which is a noxious weed native to South America that does not survive desert conditions. The other four genera are represented in our deserts and consist mainly of native species. *Matelea* and *Cynanchum* are large tropical genera with about 200 species each, but in our deserts each genus has only a single representative, one of which is rare and the other uncommon.

Most desert species of Asclepiadaceae are widespread although seldom abundant. Tending to prefer sandy or rocky washes, they do not require much water, just the occasional drink from a flash flood or a little extra runoff from a paved road.

California Desert Genera and Species

Most milkweeds in California are perennials with stiff, little-branched green upright stems. There may be no leaves or else rather large leaves, usually in pairs or whorls of three along the whole stem. Another common growth pattern is that of a twining vine with narrow opposite leaves. Flower colors are usually muted shades of purple or green on a yellowish white background, and flower clusters are somewhat rounded.

SIMPLIFIED KEY TO DESERT GENERA OF ASCLEPIADACEAE

The most obvious difference among California desert genera of Asclepiadaceae is growth form, either upright or vinelike. Other differences include leaf shape, flower color, and details of flower structure. The size and shape of filament column appendages, hoods, and horns are used to distinguish species as well as some genera. Hoods are bulbous shrouds, shaped like a hood or scoop, growing off of each filament, so they are positioned at each of the five notches between petal lobes. Sometimes there is also a horn: a curved and pointed appendage like a cow's horn, usually growing from the center of each hood and pointing toward the center of the flower.

1. Stems upright and fairly straight, not twining (11 spp) .... *Asclepias*
1' Stems twining and vinelike, often covering other plants.
  2. Flowers yellowish to orange. Petals curving inward like hoods but no hoods or horns from
    filament columns. Sepals upright between petals. Leaves linear like soft pine needles ....
    .... *Cynanchum utahense*
  2' Flowers purplish, greenish, or white. Petal lobes spreading like a star and relatively flat.
    Filament column with hoods or horns. Leaves often narrowly arrowhead-shaped.
    3. Flowers one or two at leaf axils. Filament column appendages fused in a ring with flat horns
      or lobes pointing away from the center. Leaves shaped like arrowheads .... *Matelea parvifolia*
    3' Flowers in clusters of several to many at leaf axils. Filament column with almost spherical
      hoods. Leaves like very narrow arrowheads or ovals, almost linear (2 spp) .... *Sarcostemma*

Asclepias (milkweed)—This genus provides the name for the family. There are about 100 species native to the Americas, with 15 in California, one of which is nonnative. Eleven species grow in our deserts, all of them natives. *Asclepias* may be annuals, perennials, or shrubs that either lie on the ground or stand erect. However, species in California's deserts are almost always erect perennials. Leaves are usually opposite or whorled, but a few species have alternately arranged leaves, and two species in our deserts have leaves so ephemeral you rarely see any at all. Filament column appendages include hoods and usually horns, with their positions and sizes important in distinguishing species.

Asclepias albicans (white-stemmed milkweed, wax milkweed) is one of only two desert species that is a shrub, with strong, straight, woody stems growing to about 2 m. It is usually leafless, showing off the "whitish" (*albicans*) gray-green stems. Flower clusters grow toward the tips of stems, often out of reach, but if you look closely, the rounded filament column hoods are at about the same elevation as the pistil head. This dis-tinguishes *Asclepias albicans* from *Asclepias subulata*, the other leafless milkweed, which has very long hoods, raised well above the pistil head. *Asclepias albicans* flowers have a particularly waxy look, as do older stems, accounting for the name wax milkweed. It is never abundant, but may be found in both the Sonoran and Mojave Deserts, in dry washes or on gravel slopes at 200–1,100 m.

Asclepias albicans (white-stemmed milkweed)

Asclepias asperula ssp. *asperula* (antelope horns) is a tidy, leafy milkweed about 50 cm tall. Leaves are narrow and irregularly arranged, with some in whorls of three and others alternate. It is a perennial that likes rocky places at somewhat higher elevations (1,500–2,000 m) of the Mojave Desert. *Asperula* is the diminutive form of *asper*, or "rough," meaning "a little rough," probably in reference to slight hairiness.

Sarcostemma (climbing milkweed)—There are 34 species of climbing milkweed in North America, Australia, and Africa, two of which grow in California. Both are confined to our deserts and neighboring states to the east. The twining, climbing stems of the genus are reflected in the common name. The Greek scientific name means "fleshy" (*sarco-*) "crown" (*-stemma*), referring to the bulbous filament column appendages looking like a crown around the pistil head. Petal lobes are spreading or slightly upright in a star pattern featuring the appendages, anther head, and pistil head at the center of the star.

Sarcostemma cynanchoides ssp. *hartwegii* (**climbing milkweed**) sometimes grows in a large colony, covering small trees and shrubs of a dry wash. Its opposite leaves are about 3 cm long and narrowly pointed. The leaf base is narrow too, but just wide enough to have small lobes pointed back toward the stem like barbs. Petals are purplish, usually streaked on creamy white, and bulbous filament appendages are cream-colored. The specific epithet means "resembling" (-*oides*) "cynanchum" (*cynanch-*) because the plant looks similar to species in the genus *Cynanchum*.

Examples and Uses

The most exquisite use of Asclepiadaceae is by monarch butterflies and some of their close relatives, who lay their eggs only on *Asclepias* species. Developing larvae eat milkweed leaves and sequester the cardiac glycosides without being harmed. Thus, toxin is transferred into larvae and adult butterflies, making them poisonous to birds. After getting sick on a few bright orange-and-black monarchs, birds learn to stay away. Another step in the evolutionary web has resulted in several other species of butterflies mimicking the colors of the monarch, thus minimizing predation without being toxic.

People who especially like butterflies often grow milkweed. Several species of *Asclepias* are common landscape plants needing a minimum of care.

Some aboriginal people of tropical regions learned to use Asclepiadaceae toxins on their arrowheads and spear points to improve hunting success.

A few species of *Cryptostegia* produce enough rubber from their milky sap to be useful, but they do not compete on a commercial scale with rubber trees (*Hevea brasiliensis*) in the Spurge Family (Euphorbiaceae).

Sarcostemma cynanchoides ssp. *hartwegii* (climbing milkweed)

Asclepias asperula ssp. *asperula* (antelope horns)

Encelia farinosa (brittlebush), Death Valley National Park, Mosaic Canyon

ASTERACEAE (COMPOSITAE)

SUNFLOWER FAMILY

Among all flowering plants, Asteraceae is second only to the Orchid Family (Orchidaceae) in numbers of species. The Sunflower Family is the largest family of dicots, whereas Orchidaceae species are monocots. Asteraceae have adapted splendidly to desert conditions—a habitat where orchids are rare.

Compositae is an older name for the Sunflower Family, referring to an essential feature common to nearly all members. The sunflower head that you see, resembling a single large flower, is a composite of numerous tiny flowers, specialized in various ways so that the whole group attracts pollinators and functions much the same as large single flowers. You can watch hummingbirds circle a sunflower inserting their beaks systematically into each nectar-bearing flower. Each petal radiating around the perimeter of a sunflower head belongs to a single flower, and other flowers form the central disk. Sunflowers are recognized for their habit of tracking the sun, like smart solar panels, pointing their flowers in a direction that allows the greatest absorption of warmth and maximum reflectivity of ultraviolet light to signal pollinators.

Aster, a Greek word meaning "star," refers to the radiating arrangement of flowers in a pattern resembling a star or sun.

Asteraceae Icon Features

- Numerous tiny flowers combined in a head
- Radially symmetrical flowers often clustered at the center
- Bilaterally symmetrical flowers often looking like single petals

Distinctive Features of Asteraceae

It may take close examination to identify Asteraceae when the flowers and heads are obscure, but most of them leap out at you, like daisies and sunflowers. Asteraceae members have tiny specialized flowers in a tight cluster called a **head** that usually resembles a single flower. Individual flowers in a head have no stem or pedicel but are attached directly to the **receptacle**, a round platform or dome upon which all the tiny flowers are arranged, youngest in the middle. At the base of each flower head there are sepal-like scales or bracts called **phyllaries** that form a protective **involucre**.

Individual flowers of Asteraceae are small, reduced to essentials, and variable in their parts. Flowers generally have a corolla of five fused petals (rarely four) with either radial or bilateral symmetry. There are no sepals; sepals have been modified into a hairy, scaly, or bristly **pappus** that generally remains attached to the tip of maturing ovaries after other flower parts have fallen. Asteraceae ovaries are inferior and single-seeded, so the pappus (modified sepals) attaches above the ovary and is often designed to carry its seed-containing ovary (fruit) in the wind or attach it to traveling fur or clothing. The fluffy white parachute holding a drifting dandelion seed is the pappus.

Within a head, individual flowers may have bracts, called **chaff**, growing from the receptacle along with the flowers. Chaff looks like scales or tiny stiff leaves and forms alongside developing

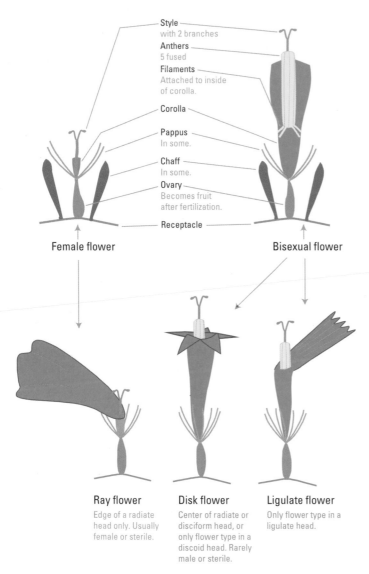

Style
with 2 branches

Anthers
5 fused

Filaments
Attached to inside
of corolla.

Corolla

Pappus
In some.

Chaff
In some.

Ovary
Becomes fruit
after fertilization.

Receptacle

Female flower

Bisexual flower

Ray flower
Edge of a radiate
head only. Usually
female or sterile.

Disk flower
Center of radiate or
disciform head, or
only flower type in a
discoid head. Rarely
male or sterile.

Ligulate flower
Only flower type in a
ligulate head.

Sunflower Family (Asteraceae) flowers are tiny and of several common types with unusual features.

ovaries rather than originating from the tips of the ovaries like the pappus. Chaff often forms an envelope around developing seeds, possibly providing protection against insect attack. Sometimes, in place of chaff, receptacles have bristles or stiff hairs among the flowers.

Flowers may be **bisexual**, female (**pistillate**), male (**staminate**), or **sterile**. Bisexual and female flowers have a single pistil with an inferior ovary bearing a single ovule, a long style split near the tip, and two stigmatic surfaces, usually on the insides or edges of style branches. Bisexual and male flowers have five (rarely four) stamens with separate filaments, each attached to the corolla tube. Anthers usually are fused, forming a tube around the style (if present). Vestigial female parts often are present in male and sterile flowers. Male flowers without any vestigial female parts sometimes have free rather than fused anthers. Frequently, female flowers consist only of a naked pistil, with the corolla as well as male parts missing.

In bisexual flowers, male and female parts often work cooperatively to avoid self-pollination. Pollen develops early in the process of flowering and coats the inside of the anther tube. As the style elongates to grow through the tube its style branches are still pressed against each other like a column split lengthwise but not separated. The stigmatic surfaces touch each other and not the pollen, but the outside surfaces of the column are like cylindrical brushes that push pollen up the tube to the outside world. After the style is fully elongated and the pollen dispersed, the style branches spread apart to expose stigmatic surfaces for fertilization by pollen from another flower.

Different combinations of flowers with various names make up heads. A major distinction is based on presence or absence of a **ligule**, an elongated tongue on one side of the corolla (the "petals" on a daisy). Flowers with a ligule are either ray flowers or ligulate flowers. There is only a single exception in our deserts, this being a two-lipped disk flower with one very elongated lip that is referred to as a ligule.

Ray flowers have a ligule formed from fusion and elongation of three petal lobes, often represented by three tiny lobes at the tip of the ligule. The other two petal lobes are absent in ray flowers, or, rarely, they occur as two teeth opposite the three-lobed ligule. Occasionally the 3:2 arrangement is reversed so the ligule has just two tiny lobes and the other three are absent. Ray flowers usually are sterile or pistillate, and they surround disk flowers in **radiate heads**, like daisies.

Ligulate flowers result from the fusion of all five petal lobes into one elongated ligule, often with five tiny lobes at its tip. Ligulate flowers are bisexual and occur only in heads with others like themselves, called **ligulate heads**. Dandelions are a familiar example.

Flowers without ligules have a corolla with five (rarely zero or four) tiny lobes, usually in radial symmetry. Radially symmetrical flowers are often clustered together in a disk surrounded by ray flowers, but even more commonly they occur in heads without ray flowers. Flowers with radial or almost-radial symmetry are called **disk flowers** if they are bisexual (or rarely male or sterile).

Radiate head - disk flowers surrounded by ray flowers.
Interior goldenbush (*Ericameria linearifolia*)

Ligulate head - all ligulate flowers.

Parry rock-pink (*Stephanomeria parryi*)

Discoid head - flowers with radial symmetry.
Rayless encelia (*Encelia frutescens*)

Discoid head - all disk flowers, but outer flowers with bilateral symmetry.
Desert pincushion (*Chaenactis fremontii*)

The most common types of heads in the Sunflower Family (Asteraceae) are made up mainly of bisexual flowers or bisexual plus ray flowers that are usually female or sterile.

When species have heads consisting of bisexual disk flowers, these are **discoid heads**. (Rarely, a few flowers in discoid heads may be male or sterile.) Discoid heads are very common in the world of Asteraceae. As described above, disk flowers frequently are surrounded by ray flowers in radiate heads—also very common.

Sometimes disk flowers have bilateral symmetry but without any elongations worthy of being defined as ligules. These are usually two-lipped disk flowers, where the five corolla lobes are divided between two lips, usually with one lip having two small lobes and the other lip having three larger lobes. In the single case of *Trixis californica*, we have a desert species where the two-lipped flower has one lip sufficiently elongated to be considered a ligule. Thistle (*Cirsium*) heads have yet another variation on disk flower corollas. They are technically bilateral, but both the tubular part and the lobes are so narrow that they might better be described as linear.

Disciform heads occur in species that have two types of radially symmetrical flowers in the same or different heads. Sometimes disciform heads are single-sex, called **pistillate heads** or **staminate heads**. These may grow on the same plant, or may

Staminate head - with several male flowers

Pistillate head - with a single female flower with bur-like fruit spines

Disciform heads
Annual bur-sage (*Ambrosia acanthicarpa*)

Disciform heads in the Sunflower Family (Asteraceae) occur on species that have two types of flowers without ligules in the same or different heads. Sometimes the sexes are separated into different heads, each having only a single type of flower.

be segregated by sex onto separate plants. More often in Asteraceae, pistillate or sterile flowers combine with staminate or bisexual flowers in disciform heads. In many cases, discoid and disciform heads are difficult to tell apart, and in our key the distinction is avoided whenever possible.

Rarely, Asteraceae flower heads contain just a single flower. In this case, the apparent flower head usually is a secondary head containing several one-flowered heads. Sometimes the pistillate heads of a species have only a single flower whereas staminate heads have several flowers.

If the flowers of Asteraceae sound variable, they are nothing compared to the plants overall. There are tiny annuals to trees (no trees in California), leaves of every description and arrangement, and a wide range of hairs, glands, spines, colors, and odors. There really are no good generalizations about Asteraceae members except for the flower heads and flowers as described above.

Similar Families

A few desert families in addition to Asteraceae have flowers in heads. One is Platanaceae, the family of sycamore trees, but these are impossible to confuse with sunflowers. Another is Polemoniaceae, the Phlox Family. In Phlox Family flower heads, each individual flower has a calyx (sepals), unlike Sunflower Family flowers. Polemoniaceae species also have three branches to their styles instead of two, their anthers are separate instead of fused, and the ovary is superior instead of inferior. In the Mint Family (Lamiaceae), the genus *Monardella* has heads that superficially resemble Asteraceae. On close inspection, however, *Monardella* is found to have a calyx for each tiny flower in the head and four stamens instead of five. Some Four O'Clock Family (Nyctaginaceae) members have flowers in heads with single-seeded ovaries that appear inferior; however, anthers are not fused.

If you think you are looking at flowers in heads but you are not sure they are Asteraceae, look carefully at individual flowers. They must not have pedicels or sepals. Anthers usually are fused into a tube, and ovaries must be inferior, containing just a single seed. Nothing but Asteraceae meets these requirements.

Family Size and Distribution

There are roughly 1,300 genera and 21,000 species of Asteraceae worldwide. They grow in virtually every habitat type but are particularly abundant in temperate, arid, and montane regions. The largest genus, *Senecio*, has about 1,500 species.

California has 207 genera of Asteraceae, most of them represented by only a handful of species. Genera with the most species in California include *Aster*, *Cirsium* (thistle), *Erigeron* (fleabane daisy), *Eriophyllum* (woolly sunflower), and *Senecio* (groundsel, ragwort, butterweed).

Our deserts have 102 genera in the Sunflower Family, of which only eight are entirely nonnative. Of the 105 California Asteraceae genera that do not occur in our deserts, about half consist entirely of nonnative species. Weedy introduced species typically find desert conditions uninhabitable, but several types of thistles and a few other species from around the globe have established themselves successfully. Several desert genera native to California have only one or a few species, often of limited distribution. A few of these are *Amphipappus*, *Atrichoseris*, *Bebbia*, *Chaetadelpha*, and *Dicoria*, just to name some from the beginning of the alphabet.

California Desert Genera and Species

Asteraceae species in California's deserts may be annuals, perennials, subshrubs, or shrubs. Vines and trees, although present in the family, do not occur in our deserts. Leaves are usually alternate or basal, but a few genera have opposite leaves. Radiate heads, with ray flowers surrounding a disk of disk flowers, are common, accounting for about half the species. Others, less conspicuous, have discoid or disciform heads, entirely lacking flowers with ligules. And finally, the smallest but most distinctive group has ligulate heads, with no disk flowers.

Yellow is a color common to both disk and ray flowers, accounting for the popular acronym DYD (damn yellow daisy), meaning that you will have to look at the details to know what it is. White and lavender are other common colors for ray flowers and for ligulate flowers. Although disk flowers are most often some shade of yellow, they can also be white, pink, rose, cream, brownish, or translucent.

The pappus is an obvious feature of many Asteraceae members that have finished blooming but still have mature fruits dispersing from spent flower heads. At this stage, the pappus fluffs up as it dries and prepares to disperse. In clusters, it takes on the look of under-the-bed dust bunnies.

SIMPLIFIED KEYS TO DESERT GENERA OF ASTERACEAE

The large number of genera in Asteraceae is best dealt with by first dividing the family into groups having consistent, readily observable features. For the most part, these groups do not represent taxonomically coherent subsets of the family and should be viewed simply as a convenience. Each group has a separate key to the genera within it. We have divided Asteraceae into seven groups based on features that are easy to distinguish without having to dissect flower heads or use magnification. If you are color blind, however, enlist the help of a friend.

KEY TO GROUPS

1. Heads ligulate, plants with milky sap .... GROUP 1
1' Heads radiate, discoid, or disciform. Plants not with milky sap.
  2. Heads radiate.
    3. Shrubs or subshrubs .... GROUP 2
    3' Annuals to perennials.
      • Ray corollas white, pink, lavender, to blue .... GROUP 3
      • Ray corollas cream, yellow, orange, or red .... GROUP 4
  2' Heads discoid or disciform.
    4. Shrubs or subshrubs .... GROUP 5
    4' Annuals to perennials.
      • Corollas white, pink, lavender, purple or blue .... GROUP 6
      • Corollas cream, yellow, orange, brown, translucent .... GROUP 7

KEYS TO GENERA BY GROUP

Several of the distinctions needed in keys are based on fine points of fruits and the pappus, which may or may not be attached. The **beak** of a fruit is an elongated, narrow extension between the seed-bearing body of the fruit and the pappus. Most fruits are not beaked, so the distinction is useful.

Pappus usually is composed of **bristles** or hairs, like a shaving brush, sometimes stiff and sometimes soft. Bristles sometimes are **barbed**, meaning that they have tiny side hooks, and sometimes are **plumose**, meaning that they have numerous tiny side branches like a feather or a plume. Pappus can also consist only of **scales,** which are broad, short, and stiff, or of **awns,** which are long, pointed, and rigid. Occasionally there are combinations of bristles, scales, or awns with varying patterns of fringing or fusion.

GROUP 1 KEY: Heads ligulate. Plants with milky sap.

1. Perennials and subshrubs.
    2. Corollas yellow (2 spp) .... *Crepis*
    2' Corollas white, pink, lavender, or blue.
       3. Corollas blue .... *Cichorium intybus*
       3' Corollas white, pink, or lavender.
          • Pappus with 5 stiff awns and many bristles. Eureka Valley .... *Chaetadelpha wheeleri*
          • Pappus with bristles only, often plumose. No awns. Widespread (4 spp) .... **Stephanomeria**
1' Annuals.
    4. Fruit beaked. Corollas white or cream to yellow.
       5. Stems under 5 cm, resting on the ground. Leaves lobed and toothed with hard whitish edges. Corollas cream to pale yellow .... *Glyptopleura marginata*
       5' Stems more than 5 cm and rising above ground level. Leaves lacking hard edges.
          • Pappus of many slender bristles, longer than fruit beak. Upper stems with large glands on the tips of hairs (2 spp) .... *Calycoseris*
          • Pappus of bristles equal to fruit beak. Plant prickly hairy .... *Lactuca serriola*
          • Pappus of plumose bristles. Plant without hairs or glands (2 spp) .... *Rafinesquia*
    4' Fruit not beaked. Corollas any color.
       6. Leaves basal and also along stems, sometimes reduced in size but not merely scale- or grasslike.
          • Fruit tapered at both ends, with veins. Pappus with 0–6 outer bristles remaining on fruit, and numerous inner bristles falling as 1. Corollas white to yellow (4 spp) .... **Malacothrix**
          • Fruit flat with many long pappus bristles. Corollas yellow (2 spp) .... *Sonchus*
          • Fruit 5-angled. Pappus plumose. Corollas pink, lavender, or white (4 spp) ... **Stephanomeria**
       6' Leaves only at the base; any above the base are scale- or grasslike.
          • Leaves broadly oval and flattened against the ground. Flowering stem tall, branching near the top. Corollas white. Pappus absent .... **Atrichoseris platyphylla**
          • Leaves pinnately lobed and lobes toothed. Phyllaries spotted or streaked with purple. Corollas pale yellow. Pappus much longer than fruit ....**Anisocoma acaulis**
          • Leaves linear, some with linear lobes. Corollas yellow, rarely white. Pappus with 2 outer bristles remaining on fruit and many inner bristles falling as 1 .... *Malacothrix californica*
          • Leaves 2–4 cm long, narrowly oval, lobed or not, scalelike above. Stems thin and highly branched. Corollas white to light pink. Fruit cylindrical, 5-ribbed. Pappus of fine bristles .... *Prenanthella exigua*

• Leaves long and narrow, grasslike, or with teeth curving toward base. Corollas yellow to reddish. Pappus of 5 scales, each notched with a bristle originating from the notch .... *Uropappus lindleyi*

GROUP 2 KEY: Heads radiate. Shrubs or subshrubs.

1. Ray corollas white or blue to lavender.
  • Ray corollas blue to lavender, rarely white, ligules up to 30 mm (3 spp) .... *Xylorhiza*
  • Ray corollas white, ligules 4–8 mm. Fruit without hairs .... *Chloracantha spinosa*
  • Ray corollas white, ligules 8–10 mm. Fruit silky hairy .... *Ericameria gilmanii*
1' Ray corollas yellow to orange or reddish.
  2. Receptacles with chaff.
    • Pappus of scales, 1–3 of which are elongated and pointed (2 spp) .... *Viguiera*
    • Pappus absent or of 2 tiny scales (4 spp) .... *Encelia*
  2' Receptacles without chaff.
    3. Fruit without pappus or hairs .... *Artemisia bigelovii*
    3' Fruit with pappus of bristles or scales, usually with hairs.
      4. Corollas all 2-lipped with longer lip 5–8 mm, shorter lip usually curled. Heads not really radiate but often appearing so. Corollas yellow. Shrubs leafy .... ***Trixis californica***
      4' Corollas of 2 types, with ray flowers (at least 1) at the perimeter of the head.
        5. Plants having a strong odor and embedded oil glands looking like small dots under the surface.
          • Ray corollas often orange, irregular in size and position. Pappus of 8–20 scales, each dividing into 5 or more bristles (2 spp) .... ***Adenophyllum***
          • Ray flowers generally 13, similarly sized, yellow. Pappus of 5 or 10 scales, each dividing into 3 sharp points .... ***Thymophylla pentachaeta***
        5' Plants sometimes aromatic and glandular but without embedded oil glands as in 5.
          6. Heads small, ray flowers sometimes only 1 or a few per head, ligules 6 mm or less.
            • Leaves small ovals, no teeth or spines. Pappus of wide twisted bristles .... ***Amphipappus fremontii***
            • Leaves long ovals, teeth with spine tips. Pappus of many bristles .. *Hazardia brickellioides*
            • Leaves threadlike. Pappus of scales with small teeth (2 spp) .... *Gutierrezia*
            • Leaves generally linear or threadlike. Pappus of fine bristles (4 spp) .... *Ericameria*
        6' Heads larger, ray flowers generally 8 or more (some 3–6), ligules over 6 mm.
          • Pappus of fine bristles (2 spp) .... *Ericameria*
          • Pappus of stiff, spreading bristles .... *Acamptopappus shockleyi*
          • Pappus of thin, barely barbed bristles (2 spp) .... *Senecio*
          • Pappus of 4–6 transparent scales. Ray corollas 3–6 and retained in reflexed position when dry .... *Psilostrophe cooperi*

GROUP 3 KEY: Heads radiate. Annuals to perennials. Ray corollas white, pink, lavender, to blue.

1. Ray corollas white, some very light pink, or with purplish lines.
  2. Disk corollas white .... *Eclipta prostrata*
  2' Disk corollas yellow.
    3. Ray flowers 150–400, white to pinkish. Ligules coiled. Streambanks ... *Erigeron philadelphicus*
    3' Ray flowers fewer than 30.
      4. Perennial from slightly woody caudex. Ray corollas slightly pinkish. **Chaetopappa ericoides**
      4' Annuals. Ray corollas very white to barely cream or ivory.
        5. Leaves to 10 cm and almost as wide, lobed and toothed. Disk corollas 4-lobed ....
        .... **Perityle emoryi**
        5' Leaves usually 2 cm or less, always narrow, some with small lobes. Disk corollas 5-lobed.
          • Plants prostrate, spreading on ground. Heads without chaff. Pappus of bristles plus scales that end as bristles, or of minute scales plus 1 plumose bristle (2 spp) .... *Monoptilon*
          • Plants spreading above ground. Heads without chaff. Pappus of scales with alternate ones ending as awns .... *Eriophyllum lanosum*
          • Plants upright. Heads with chaff between ray and disk flowers. Pappus of 10–15 awns, plumose or woolly near the base .... **Layia glandulosa**
1' Ray corollas pink, lavender, or bluish.
  6. Plants with embedded glands (dark lumps under the surface) and strong bad odor. Ray corollas generally deep pink, sometimes lighter .... **Nicolletia occidentalis**
  6' Plants without embedded glands or bad odor, but often glandular.
    7. Ray corollas lavender to bluish, ligules generally 20–30 mm extended from a tube of 4–6 mm, not coiled. Caudex woody, frequently a subshrub .... **Xylorhiza tortifolia**
    7' Ray corollas pink, lavender, to bluish, ligules 20 mm or less, frequently coiled, tube inconspicuous. Plants annual to perennial.
      • Ligules not coiled, generally 10–20 mm; if less, then pappus absent. Pappus generally of many unequal bristles. Phyllaries generally in 4–5 series, overlapping in shingle formation, some glandular but not hairy, tips curved outward or recurved (4 spp) ...... **Machaeranthera**
      • Ligules coiled, drying bluish, generally 10 mm or less, longer ones on plants over 2,000 m. Pappus of long inner bristles plus shorter outer bristles or scales often difficult to see. Phyllaries in 2 or more series, often glandular and hairy (9 spp) .... **Erigeron**
      • Ligules not coiled, under 8 mm to inconspicuous. Pappus of bristles. Phyllaries in several series, generally not glandular or hairy. Growing in wet, often alkaline places (2 spp) .... *Aster*

GROUP 4 KEY: Heads radiate. Annuals to perennials. Ray corollas cream, yellow, orange, or red.

1. Leaves mainly opposite.
  2. Receptacles without chaff.
    • Pappus absent. Ray corollas with small lobe opposite ligule. Leaves alternate above ....
    .... *Monolopia lanceolata*

- Pappus absent or of various scales or awns. Ray corollas without small lobe opposite ligule. Leaves opposite throughout (4 spp) .... *Lasthenia*
- Pappus of 20 slightly plumose bristles. Plant with embedded glands and strong spicy odor, low and spreading .... *Pectis papposa*
- Pappus of 30–40 bristles fused at the base. Ray flowers paired with phyllaries. Plant with long hairs, low growing. Some leave alternate .... *Syntrichopappus fremontii*

2' Receptacles with chaff, but sometimes only between ray flowers and disk flowers.

3. Ray flower ligules 3 mm or less.
   - Ray flowers 3–5, ligules less than 1 mm long, pale yellow. Disk flowers 1–2 ... *Madia minima*
   - Ray flowers 5–13, ligules 2–3 mm, leathery, yellow, 2-lobed. Disk flowers many .... *Sanvitalia abertii*

3' Ray flower ligules 8 mm or more.
   - Pappus absent. Leaves 2–6 cm, narrow, no teeth, tapered to stem .... *Heliomeris multiflora*
   - Pappus absent. Leaves long-triangular, irregular teeth .... *Verbesina encelioides*
   - Pappus of 2–4 awns. Disk corollas yellow throughout .... *Bidens laevis*
   - Pappus of scales. Disk flower corolla lobes red .... *Helianthus ciliaris*

1' Leaves alternate throughout, all basal, or alternate and basal.

4. Receptacles with chaff, sometimes only between ray flowers and disk flowers.

5. Pappus of awns or bristles, sometimes also with scales.
   - Pappus of disk flowers either awns or bristles. Leaves narrow, spread along stem, some lobed (2 spp) .... *Layia*
   - Pappus of 2 awns plus short scales. Ray ligules 2–5 cm. Leaves broad, nearly all basal, no teeth (2 spp) .... *Enceliopsis*
   - Pappus of 2 awns. Ray ligules 1–2 cm. Leaves broad, basal plus several on stem, some toothed .... *Geraea canescens*

5' Pappus absent or of scales; no bristles or awns but scales may be pointed.
   - Disk flower corolla lobes dark red to purple. Fruit flattened, not beaked. Stems and/or leaves with bristles or hairs (2 spp) .... *Helianthus*
   - Disk flower corollas pale yellow. Fruit beaked. Stems with bristles or hairs, and often with glands (2 spp) .... *Hemizonia*
   - Disk flower corollas deep yellow. Fruit flattened, not beaked. Stems and leaves without hairs, glands, or bristles (3 spp) .... *Coreopsis*

4' Receptacles without chaff.

6. Pappus absent or of scales or awns (no bristles).

7. Pappus of awns or scales with awned or pointed tips.
   - Pappus of 2–3 awns. Fruit without hairs. Phyllaries in 4–5 series, shinglelike, without hairs, tips often reflexed (3 spp) .... *Grindelia*
   - Pappus of several (usually 5) awn-tipped or pointed scales. Fruit hairy. Phyllaries in 2–3 series, usually hairy (3 spp) .... *Hymenoxys*

7' Pappus absent or of short or rounded scales.

8. Ray corollas reflexed and drying papery, not falling with age.
   - Pappus absent. Plants annual. Ray flowers generally many; if few, then pale yellow (3

spp) .... *Baileya*

• Pappus of 4–6 scales. Plants perennial to subshrubs. Ray flowers 3–6, deep yellow ....
.... *Psilostrophe cooperi*

8' Ray corollas falling in age, not retained in reflexed position.

• Ray corollas 30–60, red, linear, 6–10 mm long, and hairy .... *Hulsea heterochroma*

• Ray corollas 9–32, yellow, 12–18 mm long. Leaves all basal .... *Hulsea vestita*

• Ray corollas 5–10, yellow, 2–10 mm long. Leaves basal and along stem, entire to barely lobed. Plants low growing, under 30 cm (2 spp) .... *Eriophyllum*

• Ray corollas 10–20, yellow, 5–8 mm long. Leaves basal and along stem, deeply lobed with lobes lobed again. Plants tall, usually over 50 cm .... *Bahia dissecta*

6' Pappus of bristles, sometimes also scales.

9. Plants requiring moist or alkaline habitats.

• Plants in damp sand, under 700 m, eastern Sonoran Desert. Leaves 1–3 cm. Ray flowers many, ligules linear, 1.5–2 mm long .... *Pulicaria paludosa*

• Plants on stream banks or in meadows of northern desert mountains. Lower leaves to 25 cm. Ray flowers 3–13, ligules 1–2.5 mm long .... *Solidago confinis*

• Plants in saline meadows or dry alkaline flats, Mojave Desert. Lower leaves to 36 cm. Ray flowers 7–28, ligules 5–10 mm .... *Pyrrocoma racemosa*

9' Plants not requiring moist or alkaline habitats.

10. Leaves narrow, not lobed or toothed.

• Heads in flat-topped clusters, 4–9 ray flowers and 30–50 disk flowers per head (2 spp) *Heterotheca*

• Heads in flat-topped clusters, 2–3 ray flowers and 2–4 disk flowers per head ....
.... *Petradoria pumila*

• Heads solitary at branch tips, 6–15 ray flowers and 25–50 disk flowers per head ....
.... *Stenotus acaulis*

10' Leaves lobed to deeply dissected, some with basal lobes clasping stem, blades generally with small teeth. Phyllaries in 3–6 series of differing lengths.

• Leaves lobed at base, variable up tall stem, generally without petioles, blade bases clasping stem (3 spp) .... *Heterotheca*

• Leaves pinnately lobed. Phyllaries in 4–6 series, overlapping in shingle formation (2 spp) *Machaeranthera*

• Leaves slightly lobed or finely pinnately divided. Main phyllaries edge to edge in 1 series, with a much smaller series below (3 spp) .... *Senecio*

• Leaves 3-lobed at tip. Some leaves may lack lobes. Phyllaries in 1 series, each folded around the base of a ray flower .... *Syntrichopappus fremontii*

GROUP 5 KEY: Heads discoid or disciform. Shrubs or subshrubs.

1. Sexes in different heads, sometimes on different plants.

• Sexes on different plants, only pistillate or only staminate heads on a given plant. Pistillate heads with at least 8 flowers each (5 spp) .... *Baccharis*

- Sexes on same plant. Pistillate heads with a single flower. Fruits surrounded by winged chaff scales like the brim on a hat (2 spp) .... *Hymenoclea*
- Sexes on same plant. Pistillate heads with 1–2 flowers. Fruits surrounded by spines forming a bur (4 spp) .... *Ambrosia*

1' Sexes in same heads, either bisexual flowers alone in heads or various flowers combined in heads.

  2. Receptacles with chaff.

- Leaves 5–15 cm and narrow, sometimes deeply lobed with lobes narrow. Corolla absent in female flowers, translucent in male flowers. Fruit covered with long soft hairs .... *Iva acerosa*
- Leaves 1–2 cm, oval. Flowers all bisexual, corollas yellow. Pappus of 2 narrow scales or zero ... . *Encelia frutescens*
- Leaves generally absent, or 1–3 cm and narrow, sometimes few lobed. Flowers all bisexual, corollas yellow. Pappus of plumose bristles .... *Bebbia juncea*

  2' Receptacles without chaff.

    3. Genera with easy-to-spot unusual features.

- Heads of 2-lipped disk flowers with 1 lip elongated into a ligule 5–8 mm long, shorter lip about 5 mm and coiled, corollas yellow. Shrubs densely leafy. Leaves with wide petiole, blades 2–11 cm, usually with fine teeth .... *Trixis californica*
- Heads single-flowered, clustered, and surrounded by broad spine-toothed bracts 1–2 cm long looking like leaves. Leaves below narrow, spine-tipped, sparse ...... *Hecastocleis shockleyi*
- Plant covered with embedded glands having a bad odor. Glands like dark oval-shaped warts dotting phyllaries, leaves, and stems .... *Porophyllum gracile*
- Leaves looking like tiny spears, with long petioles for shafts and shorter blades (about 5 mm) coarsely toothed like a spear head .... *Pleurocoronis pleuriseta*

  3' Plants not as in 3.

    4. Fruits without pappus, definitely no bristles.

- Subshrub with simple hairs. Heads discoid. Corollas 4-lobed (2 spp) .... *Perityle*
- Subshrub silky with long forked hairs, 1 side longer than the other. Heads disciform with few pistillate flowers around more disk flowers. Flowers clearly visible, not hidden by phyllaries .... *Sphaeromeria cana*
- Shrub with hairs simple or T-shaped. Heads discoid or disciform. Flowers hidden by phyllaries and inconspicuous. Leaves often 3-lobed at tip (4 spp) .... *Artemisia*

  4' Fruits with a pappus of bristles, stiff or soft, various numbers, occasionally with long narrow scales.

    5. Leaves broad, oval to spoon-shaped, and often toothed or notched at the tip.

- Corollas white to marked reddish. Leaves usually broad and toothed (10 spp) ... *Brickellia*
- Corollas pink to rose. Leaves narrow ovals, 1–4 cm long, entire, crowded throughout stems .... *Pluchea sericea*
- Corollas yellow. Leaves spoon-shaped and often notched at the tip ... *Ericameria cuneata*
- Corollas yellow. Leaves oval and often toothed .... *Isocoma acradenia*

    5' Leaves scalelike or linear (needle- or threadlike) and without teeth.

      6. Leaves of at least upper stems scalelike.

- Heads solitary. Leaves linear below, scalelike above .... *Machaeranthera carnosa*

• Heads clustered. Leaves of flowering stems scalelike .... *Lepidospartum squamatum*

 6' Leaves longer than scalelike: linear, needlelike or threadlike, and without teeth.

  7. Phyllaries in 1 series, side to side in 1 circle.

    • Heads with 12–21 yellow flowers and with leafy bracts. Phyllaries 9–18. Shrubs covered with leaves that look like fir-tree needles .... *Peucephyllum schottii*

    • Heads with 4–8 cream to yellow flowers. Phyllaries 4–5. Plants often spiny and leafless (5 spp) .... *Tetradymia*

 7' Phyllaries in several series and overlapping, similar to shingles on a roof.

  8. Corollas white, some red tinged.

    • Pappus of stiff bristles. Fruit 10-ribbed, hairy. Leaves generally 3–8 cm (3 spp) ....*Brickellia*

    • Pappus of soft bristles. Fruit 5-ribbed, hairy. Leaves 2–3 cm... *Chrysothamnus albidus*

  8' Corollas pale yellow to yellow.

    • Involucre hemispheric to spheric. Phyllaries oval in 2–3 series, bases cream, tips green, edges papery. Pappus of 20–25 broad stiff bristles spreading widely apart. Leaves less than 15 mm long .... *Acamptopappus sphaerocephalus*

    • Involucre cylindric. Phyllaries grading to bracts below. Plants broomlike. Leaves 2–3 cm long, thread- or needlelike. Corollas pale yellow .... *Lepidospartum latisquamum*

    • Involucre cylindric. Phyllaries ridged along center back, no bracts below. Leaves usually wider than threadlike. Corollas yellow (7 spp) .... *Chrysothamnus*

GROUP 6 KEY: Heads discoid or disciform. Annuals to perennials. Corollas white, pink, lavender, purple, to blue.

1. Leaves and stems with sharp spines.

  • Heads discoid. Leaves along stem sparsely or not hairy. Pappus bristles flat and barbed............. .... *Carduus nutans*

  • Heads single-sex, on separate plants. Leaves sparsely or not hairy. Pappus bristles plumose .... *Cirsium arvense*

  • Heads disciform. Leaves with long wavy hairs. Pappus bristles plumose (4 spp) .... *Cirsium*

1' Leaves and stems without spines, unarmed.

  2. Leaves basal or alternate all along stems.

    • Corolla white to pinkish. Leaves generally pinnately divided 1–5 times (6 spp) ..... *Chaenactis*

    • Corollas purple. Leaves oval, toothed.. *Pluchea odorata*

  2' Leaves opposite below or throughout, not pinnately divided.

   3. Plants perennial with a woody caudex... *Ageratina herbacea*

   3' Plants annual.

    • Plants hairy. Leaves 2–12 cm, narrow ovals, no lobes or teeth .... *Palafoxia arida*

    • Plants hairy. Leaves 3–5 cm, triangular to round, some with small lobes... *Dicoria canescens*

    • Plants without hairs. Leaves under 5 cm, linear, rarely lobed or toothed .... *Malperia tenuis*

GROUP 7 KEY: Heads discoid or disciform. Annuals to perennials. Corollas cream, yellow, orange, brown, or translucent.

1. Phyllaries well armed with long sharp spines.
    • Leaves not spine-tipped. Disk corollas 10–12 mm (2 spp) .... *Centaurea*
    • Leaves spine-tipped. Disk corollas 20 mm. Receptacle with bristles. .... *Cnicus benedictus*
    • Leaves spine-tipped. Disk corollas 25–35 mm. Receptacle with chaff .... *Carthamus baeticus*
1' Phyllaries not armed, sometimes absent, often hairy.
    2. Sexes in different heads on the same plants. Pistillate heads forming fruits with burs.
        • Staminate heads in tight clusters. Fruits covered with velcro-like hooks ....
        .... *Xanthium strumarium*
        • Staminate heads in loose elongated clusters, nodding. Fruits with several to 30 straight spines
        .... *Ambrosia acanthicarpa*
    2' Sexes in same heads. Fruits not forming burs.
        3. Receptacles with chaff.
            4. Phyllaries partly fused. Heads single in leaf axils, nodding. Chaff scattered .... **Iva axillaris**
            4' Phyllaries absent or vestigial. Heads in small clusters surrounded by leafy bracts. Chaff scales folded around outer flower ovaries.
                • Chaff scales with spines recurved or hooked at tips .... *Ancistrocarphus filagineus*
                • Chaff without hooked spines. Disk flowers bisexual. Pappus of 11–23 bristles. Corolla lobes 4 or 5 (3 spp) .... *Filago*
                • Chaff without hooked spines. Disk flowers staminate, ovary vestigial. Pappus of 1–8 bristles. Corolla lobes 5 (5 spp) .... *Stylocline*
        3' Receptacles without chaff or bristles.
            5. Pappus absent or of scales only, no bristles.
                6. Pappus absent.
                    • Heads bulbous, displaying numerous 4-lobed disk flowers .... *Chamomilla occidentalis*
                    • Heads small, flowers few, obscured by phyllaries, disk corollas 5-lobed (3 spp) ....
                    .... *Artemisia*
                6' Pappus of scales, some less than 1 mm.
                    • Annuals, 1–8 cm. Leaves woolly, wedge-shaped, 3-lobed at tip (2 spp) .... **Eriophyllum**
                    • Perennials, 30–70 cm. Leaves woolly, mostly basal, dissected into linear lobes ....
                    .... *Hymenopappus filifolius*
                    • Annuals, 5–25 cm. Leaves with short stiff hairs, basal and along stems, lower opposite and dissected into threadlike lobes .... *Schkuhria multiflora*
                    • Annuals, under 50 cm. Leaves woolly, generally pinnately lobed ....
                    .... *Chaenactis glabriuscula*
            5' Pappus of bristles or scales dissected into bristles.
                7. Corollas cream to pale yellow, some with reddish tips.
                    • Annual, to 2 m, single-stemmed, branching above among numerous small heads. Plants with short stiff hairs or longer soft hairs, if soft, then leaves at least partially lobed. Heads may have vestigial or inconspicuous ray flowers (2 spp) .... *Conyza*

- Annual to perennial, to 70 cm, single- to several-stemmed, branched above or below. Plants covered with woolly hairs. Leaves various narrow shapes, not lobed (4 spp) .... *Gnaphalium*

7' Corollas yellow.

   8. Corollas of outer flowers 2-lipped, longer lobes extending outward.

      - Phyllaries in 4–9 series, shingle formation, margins glandular .... *Lessingia lemmonii*

      - Phyllaries in 2 nearly equal series, hairy .... *Tricoptilium incisum*

   8' Corollas all with radial symmetry, not 2-lipped.

      9. Leaves hairy.

         - Plants low, much branched, spreading, and rounded. Leaves short-hairy, almost round. Heads at leaf level (2 spp) .... *Psathyrotes*

         - Leaves mostly basal, long and long-hairy. Heads single to several on stems above leaves. Heads sometimes have vestigial ray flowers .... *Erigeron aphanactis*

      9' Leaves not hairy.

         - Phyllaries in 1 main upright series, plus much smaller spreading phyllaries below. Heads of more than 10 disk flowers; ray flowers sometimes present, inconspicuous or not (2 spp) .... *Senecio*

         - Phyllaries in several series of different lengths, overlapping. Heads of 4–7 disk flowers .... *Chrysothamnus gramineus*

Acamptopappus shockleyi (Shockley goldenhead)

Acamptopappus (**goldenhead**)—This genus has a Greek name describing its pappus as "unbending" or "stiff" (*acampto-*, usually *acantho-*). A dense coating of long hairs on fruits is also distinctive but not unique to the genus. Receptacles are deeply pitted where individual flowers attach, with projections in-between, but there is no chaff. There are just two species of goldenhead, both in our deserts: one with ray flowers and one without.

Acamptopappus shockleyi (**Shockley goldenhead**) is a subshrub less than 40 cm tall with numerous stems that terminate in a radiate head of yellow flowers, with 5–14 ray flowers surrounding 30–80 disk flowers. *Acamptopappus shockleyi* sometimes grows in abundance in small areas on stony slopes and ridges of the Mojave Desert at elevations of 500–2,000 m. It is named for William H. Shockley (1855–1925), a mining engineer who collected plants in eastern California.

Acamptopappus sphaerocephalus (**goldenhead**) is a heavily branched subshrub up to 1 m tall. Its numerous heads usually grow in clusters near branch tips. The Greek "spherical" (*sphaero-*) "headed" (*cephalus*) refers to the absence of ray flowers and the shape of the head, which contains 13–27 disk

Acamptopappus sphaerocephalus (goldenhead)

Ambrosia (ragweed, bur-sage)—This genus of numerous poorly defined species, ranging from annuals to shrubs, grows all over North America. Several species are known for their allergy-producing pollen. The Greek genus name is a mythological food for the gods, based on aromatic qualities. Bur-sage best describes our four California desert species, which have one to two female flowers without corollas in heads by themselves, with the chaff scales growing into spines around the fruit(s). Male flowers have a tiny yellow or translucent corolla, and many grow in single-sex heads. Staminate heads are on spikes above smaller clusters of pistillate heads.

Ambrosia dumosa (burrobush) is one of the most common plants found below 1,600 m throughout our deserts. It is a shrub with numerous white-hairy branches and leaves. Leaves are usually 1–2 cm long and irregularly divided into deep lobes and deep teeth or smaller lobes. The whole surface is crinkled or wavy and light gray-green. Although

flowers, golden in color. There are two inter-grading varieties covering elevations of 60–2,200 m, mainly in the Mojave Desert.

Adenophyllum **(dyssodia)**—This genus of 10 species has two species of subshrubs in our deserts. They are characterized as "gland" (*adeno-*) "leaf" (*phyllum*) for the oil glands embedded in leaves and phyllaries. The common name (former genus name), dyssodia, refers to the very strong unpleasant odor produced by the glands. Ray corollas are few, oddly spaced, uneven in size, and occasionally absent. Both ray and disk flowers are deep yellow to orange or even reddish. There is no chaff. Pappus consists of scales that terminate in long bristles.

Adenophyllum cooperi **(Cooper dyssodia)** has simple coarsely toothed or shallowly lobed leaves and 15–20 pappus scales, whereas its cousin *Adenophyllum porophylloides* has deeply lobed and toothed leaves and 8–12 pappus scales. *Adenophyllum cooperi* grows in sandy parts of the Mojave Desert, including roadsides, from 600 to 1,550 m.

Adenophyllum cooperi (Cooper dyssodia)

individual flower heads are unnoticeable, spikes of staminate heads with clusters of fruits below on a dense gray-green shrub make this important plant easy to spot. The Latin specific epithet refers to the bushy look of the species, and perhaps hungry burros eat it, providing a common name.

Ambrosia ilicifolia (**holly-leaved bur-sage**) is the "holly" (*ilici-*) "leaved" (*folia*) species. Stems have short stiff hairs, and leaves are spine-tipped. It grows in sandy washes and rocky canyons of the southern deserts, under 500 m.

Amphipappus (**chaff-bush**)—This is a single-species genus with a "double" (*amphi-*) pappus. It is the spiny quality of leafless stems that accounts for the common name: the flower heads have no chaff.

Amphipappus fremontii (**eytelia**) is a shrub with numerous thin but stiff branches. Its small flower heads have just one or two ray flowers, at first appearing defective or partially eaten. Another plant with this odd feature of its small flower heads is *Gutierrezia microcephala* (sticky snakeweed), but *Amphipappus fremontii* can be easily distinguished by its oval leaves. It grows on rocky slopes and in canyons of the eastern Mojave Desert under 1,600 m. The species is named for John C. Frémont, a western explorer and politician with an interest in botany. Several species and a genus are named for him. The common name honors Carl Eytel (1873–1927), a desert lover and friend to botanists.

Anisocoma (**scale-bud**)—This is another single-species genus named for its characteristic pappus, called "unequal" (*aniso-*) "hair-tuft" (*coma*) because some of the plumose

Ambrosia dumosa (burrobush)

Ambrosia ilicifolia (holly-leaved bur-sage)

pappus bristles are much longer than others.

Anisocoma acaulis (**scale-bud**) further describes the species as being "with-

Amphipappus fremontii (eytelia)

out" (*a-*) a "stem" (*caulis*). The leaves grow just at sand level, and a single ligulate head tops a smooth bare flower stalk up to 20 cm long. Corollas are pale yellow, and the phyllaries that surround them have papery, translucent edges and reddish spots near their tips. In buds, phyllaries look scaly, providing a common name. Scale-buds are annuals with milky sap. They often grow in colonies in sandy washes or on gravelly flats at 600–2,400 m in both deserts.

Atrichoseris (**tobacco-weed, gravel-ghost**)—This is another single-species genus named for its pappus—in this case, its absent pappus. "Without" (*a-*) "hair" (*tricho-*) is added to *seris*, the Greek word for plants like chicory, which is sometimes used as a substitute for tobacco.

Anisocoma acaulis (scale bud)

Atrichoseris platyphylla (**tobacco-weed, gravel-ghost**) has "flat" (*platy-*) "leaves" (*phylla*), widely oval in shape and hugging the ground. Its ligulate heads of white flowers appear as ghosts over the gravel because the stems that support them are thin and tall, with the leaves often over 1 m away. Gravel-ghost grows in gravelly valleys and washes under 1,400 m.

Baccharis (**baccharis**)— This genus of as many as 400 species, all in the Western Hemisphere, is named for the Latin god of wine but has nothing to do with grapes. There are nine

Atrichoseris platyphylla (gravel-ghost) in Death Valley National Park, Telescope Peak behind

species of *Baccharis* in California, five of which occur in our deserts, often in association with water. *Baccharis* may be perennials to shrubs, all with the unusual feature of having the sexes segregated onto different plants. All desert species are shrubs with heads that contain either one or two kinds of flowers but only one sex. Flowers on male shrubs retain a vestigial ovary with a small pappus, but pollen will assure you that male parts are functional. Female shrubs have flowers with a corolla reduced to a thread and a cylindrical ribbed ovary with a pappus of many bristles that extend well past other flower parts. Either type of head sometimes includes sterile flowers.

Baccharis salicifolia (mule fat), female flowers (left) and male flowers (right)

Baccharis salicifolia (**mule fat, seep-willow, water-wally**) is widespread in California and extends to South America. It is the most attractive of the desert *Baccharis*, "willow" (*salci-*) "leaved" (*folia*), with large clusters of pink flower heads. Shrubs grow to 3 m or more, usually along streams or ditches under 1,250 m.

Baileya (**desert-marigold**)—This genus of four species was named for Jacob Whitman Bailey, a microscopist born in 1911. Three of the species occur in California, including our deserts. They are generally annuals, although some may live longer than a year, and they are covered with soft hairs so that they appear gray-green and almost fuzzy. Species have varying numbers of ray flowers and disk flowers, both in shades of yellow. Receptacles sometimes have narrow chaff, but usually there is none. Fruits are ribbed and sometimes angled, without pappus.

Baileya multiradiata var. *multiradiata* (**wild marigold, desert-marigold**) has 50–60 ray flowers in several layers surrounding many disk flowers, hence its name "many" (*multi-*) "rayed" (*-radiata*). This species is distinguished from *Baileya pleniradiata* by having most of its leaves near the base of the stems and not near the solitary flower heads. Wild marigold grows on sandy slopes and washes or roadsides, mainly in the eastern Mojave Desert at elevations of 600–1,600 m.

Baileya multiradiata var. *multiradiata* (wild marigold)

Baileya pauciradiata (**lax flower**), though "few" (*pauci-*) "rayed" (*-radiata*), is still spectacular—if perhaps a bit more subtle than its cousins. Corollas are pale lemon yellow and herbage is thickly hairy. Lax flower likes loose sand under 1,100 m.

Bebbia (sweetbush)—There are two species in this genus named for Michael Schuck Bebb (1833–1895), an American botanist who specialized in willows. Not much about *Bebbia* looks like willows (Salicaceae), however, at least in the one species that grows in California.

Bebbia juncea var. *aspera* (sweetbush, chuckawalla's delight) has numerous green stems that are "reed-" or "rush-like," according to the specific epithet based on the Latin word *junc.* A multitude of stems form a dense, almost spherical shrub or subshrub about 1 m tall. Stems are brittle and usually have short stiff hairs that feel like sandpaper. Many of the stems terminate in a small head of yellow to orange disk flowers that are delicacies to chuckawallas (a type of lizard). Fruits are dark and angled with ascending white hairs and a pappus of 15–30 slightly plumose bristles up to 1 cm long. Sweetbush is common throughout the deserts under 1,500 m, particularly in dry washes.

Brickellia (brickellbush)—The genus is named for John Brickell (1749–1809), an Irish physician who settled in Savannah, Georgia. There are 15 species of brickellbush in California, 13 of which occur in our deserts. All desert species are shrubs except for one subshrub, although the genus as a whole includes perennials and a few annuals. Most species in our deserts have fairly broad, toothed, thick-veined leaves dotted with resin. Small heads of disk flowers without chaff are usually clustered near branch tips. Corollas generally are white or tinged reddish. Fruits are cylindrical with 10 ribs and hairy. The pappus consists of many stiff, straight bristles.

Brickellia arguta (spear-leaved brickellbush) grows in rocky places under 1,500 m. It is branched in a zigzag manner to a height of 30–40 cm and has leathery pointed and toothed leaves up to 2 cm long and 1 cm wide. The Latin *arguta*, meaning "sharp" or "pungent," may refer to the leaves or the pointed

Baileya pauciradiata (lax flower)

Bebia juncea var. *aspera* (sweetbush)

phyllaries. Heads contain 40–55 undistinguished disk flowers.

Brickellia incana (**woolly brickellbush**) grows in sandy places under 1,700 m, mostly in the eastern deserts. It is densely branched into a sphere up to 1 m high, and it is nearly white (or "gray" as in the Latin *incana*) from long dense hairs on stems, leaves, and phyllaries. Leaves are broad, pointed ovals usually about 1 cm long and barely scalloped or toothed. Each head contains about 60 disk flowers, which appear greenish to maroon.

Calycoseris (**tack-stem**)—There are two species of tack-stem, both growing in California's deserts. The common name refers to stalked glands in the shape of tacks that grow along the upper stems. Each gland is like the head of a tack,

Brickellia arguta (spear-leaved brickellbush)

Brickellia incana (woolly brickellbush)

borne on the end of a hair or tiny stalk. "Cuplike" (*calyco-*) "chicory" (*seris*) suggests that this plant is related to other chicories, with ligulate heads and milky sap. Fruits have a short beak terminating in a white pappus of many slim bristles that fall as one.

Calycoseris parryi (**yellow tack-stem**) is named for Dr. C. C. Parry (1823–1890), an American botanist who collected extensively in California and is commemorated in many species names. Yellow tack-stem has yellow corollas, while white tack-stem (*Calycoseris wrightii*) has white corollas. They are equally showy annuals with telltale glands and ligulate heads. *Calycoseris parryi* is broadly distributed in both deserts on sandy to gravelly soils under 2,000 m.

Chaenactis (**pincushion**)—There are 18 species of *Chaenactis* (meaning "gaping" [*chaen*] "ray" [*actis*]), all in western North America; 13 are found in California, of which six occur in our deserts. They may be annuals to subshrubs with alternate or basal leaves that are deeply lobed, sometimes with lobes further dissected. Heads are made up entirely of disk flowers, but sometimes the outermost flowers have enlarged outer corolla lobes looking somewhat like very short fat ligules. Desert species are all annuals or perennials. Most species have white to pinkish corollas, but one, *Chaenactis glabriuscula*, has yellow flowers.

Chaenactis fremontii (**desert pincushion, Fremont pincushion**) is a common species under 1,600 m, growing in open sand or gravel in

Calycoseris parryi (yellow tack-stem)

both deserts. It is an annual, usually with several stems branched below the middle and basal leaves that are somewhat fleshy. The outer corollas in each head are bilateral, with three lobes enlarged.

Chaetopappa—This genus has the Greek name meaning "bristle" (*chaeto-*) "pappus" (*pappa*), but a few of the approximately 10 species do not fit this description. *Chaetopappa* may be annuals to subshrubs having radiate heads without chaff. Ray flowers have white, pinkish, or bluish ligules, and disk corollas are yellow. Only one species is found in our deserts.

Chaenactis fremontii (desert pincushion)

Chaetopappa ericoides (white aster) is a perennial, only about 10 cm tall, with stems growing from a slightly woody caudex at ground level. Stems and leaves have stiff hairs and are usually glandular. The 12–21 ray flowers have white to pinkish corollas. Fruits are rounded and hairy with a pappus of minutely barbed bristles. *Chaetopappus ericoides* grows on dry slopes of desert mountains from 1,300 to 2,900 m, even though its specific epithet means "resembling" (*-oides*) "heath" (*erica*) such as you would find on a Scottish moor.

Chaetopappa ericoides (white aster)

Chrysothamnus (rabbitbrush)—The Greek words "gold" (*chryso-*) "shrub" (*thamnus*) describe this genus in bloom, when most species are densely coated with small but distinctly gold or yellow flowers in numerous tiny heads. Each corolla is radially symmetrical, with spreading lobes. Style branches also are elongated and spreading, extending well past the corolla. Fruits generally are long and cylindrical with five ribs and a pappus of many bristles. There are perennials, subshrubs, and shrubs, often heavily branched and growing densely enough to please rabbits, hence the common name, rabbitbrush. However, most humans are not pleased when they discover the strong allergenic qualities of the pollen. There are 16 species of *Chrysothamnus* in western North America; 10 of them occur in California, all but one of which grow in our deserts.

Chrysothamnus nauseosus (rubber rabbitbrush) is the most widespread of the *Chrysothamnus* species, and several of its 22 subspecies grow in the Mojave Desert, including desert mountains. The specific epithet derives from the strong allergic response people often have to shrubs in flower. Plants also have a strong odor, but it generally is not considered unpleasant. Shrubs, sometimes well over 1 m in height, are gray-green from long silvery stems. Leaves, also hairy, are slim and up to 7 cm long, but frequently they are almost entirely absent. There are generally five flowers per head, with heads growing in large flat-topped or domed clusters that can be seen from a distance when they bloom in the fall.

Chrysothamnus nauseosus (rubber rabbitbrush)

Chrysothamnus viscidiflorus (yellow rabbitbrush, sticky-leaved rabbitbrush) is a shrub growing to about 1 m, with woody branches that turn white with age. Otherwise the stems and leaves are strongly green, and the long narrow leaves are usually twisted and sticky. Phyllaries are green and sticky too, resulting in the Latin name *viscidi-* (sticky) *-florus* (flowered). Each head has 3–13 flowers, and heads are in dense clusters. *Chrysothamnus viscidiflorus* grows at 900–4,000 m in our desert mountains.

Cirsium (thistle)—*Cirsium* is the Greek word for thistle, and there are plenty of them, roughly 200 species, throughout North America and Eurasia. Several of the 24 species occurring in California originated elsewhere, but only one of our five desert species is nonnative. Thistles often are biennial, with just a cluster of leaves at ground level the first year, followed by a rapidly growing, tall flowering stalk the second year. Leaves are long and wavy-edged with irregular lobes and teeth that are well endowed with sharp spines. Many disk flowers generally are crowded into large heads surrounded by numerous phyllaries, the outer ones spine-tipped. Corollas are long, very thin, and usually white to pink or purplish.

Chrysothamnus viscidiflorus (yellow rabbitbrush)

Cirsium mohavense (**Mojave thistle**), growing in the Mojave Desert (*mohavense*) at 400–2,800 m near springs and seeps, is one of the most spectacular thistles, with flower heads glowing pink with highlights of yellow and rose.

Cirsium neomexicanum (**desert thistle**) grows in the eastern Mojave Desert and points east, including New Mexico (*neomexicanum*), at 800–2,100 m. Desert thistle is tall and handsome in flower, but stems and involucres are covered with cobwebby hairs so that close inspection makes you want to get out a dust cloth. Corollas are white to very pale pink or lavender.

Coreopsis (**tickseed**)—The Greek genus name, reflected in the common name, refers to the fruits and their "appearance of"

Cirsium mohavense (Mojave thistle)

(-*opsis*) "a bug or tick" (*core*-). Indeed, broad and compressed in shape, the fruit without a pappus, or (in some cases) with two awns or scales, does look somewhat like a flat round bug with antennae. *Coreopsis* is a genus of about 114 species found mainly in the Americas and Africa. California has 11 species, three of which are nonnative, while three others live in our deserts. Although the genus has annuals, perennials, and shrubs, our desert species are all annuals. Heads are radiate with chaff, and phyllaries are in two distinct circles, with the inner ones thin and generally upright and the outer ones thick and spreading widely. Both ray and disk flowers are yellow in desert species, and the shade of yellow is toward gold. Ray flowers may be either fertile or sterile, depending on the species. Leaves and stems are

Cirsium neomexicanum (**desert thistle**), Kingston Mountains

Coreopsis californica var. *californica* (California tickseed)

thin, completely without hairs or glands, and often look shiny and reddish.

Coreopsis californica var. *californica* (**California tickseed**) is common on our low desert plains and washes after good spring rains. It usually has many stems, with threadlike leaves clustered almost entirely at the base. There are 5–12 fertile ray flowers and 10–30 disk flowers.

Dicoria—This unassuming genus of three species is named "two" (*di-*) "bugs" (*coria*) because there are usually two female corolla-free flowers in a head, and their compressed winged fruits without a pappus vaguely resemble bugs. Heads have 5–20 male flowers with small greenish to purple corollas, and both sexes usually occur in the same heads.

Dicoria canescens (**desert dicoria**) is named for having an "ash-colored" (*canescens*) appearance due to dense whitish hairs. Lower leaves are opposite, 3–5 cm long and almost as broad, with a pointed tip. Upper leaves are smaller, round, and arranged alternately. The tiny heads are spread along branch tips and generally hang face down. Plants can be almost 1 m tall but are hardly noticeable. They like sandy places under 1,300 m.

Encelia (**brittlebush**)—This genus, named after Christopher Encel, a sixteenth-century botanist who wrote about oak galls, consists of 13 freely hybridizing species native to the western Americas. Five species plus one recognized hybrid grow in California, all but one of which occur in our deserts. *Encelia* are shrubs, densely branched from the base, growing up to 1.5 m. The leaves, which are fairly large and broad, fall off during extended dry periods. Most species have quantities of spectacular radiate heads with sterile yellow ray flowers and lots of chaff. Disk flowers are numerous and either yellow or brownish. Their fruits are strongly flattened and have no pappus or just two tiny scales.

Encelia farinosa (**brittlebush, incienso**) grows all over stony hillsides of the desert below 1,000 m. It has 11–21 ray flowers per head and several heads per stem. In full bloom it can color the landscape with its rounded gray bushes topped in yellow. *Farinosa* is Latin for "mealy" and refers to the granular surface of the leaves. This is the only species with several flower heads

Dicoria canescens (**desert dicoria**)

Encelia farinosa (brittlebush)

Encelia frutescens (rayless brittlebush)

on a loosely branched naked flower stalk. Other species have a single head on an unbranched naked flower stalk.

Encelia frutescens (**rayless brittlebush**) is common in washes and on flats or slopes under 800 m. Its specific epithet could allude to the fact that it lacks ray flowers and its rounded heads of disk flowers look like "fruits." However, in Latin *frutescens* means "shrublike," a quality true of all *Encelia*.

Enceliopsis (**sunray**)—This genus's name means that it looks like *Encelia*, and while this is true of the flowers, the plants are much smaller and are perennials rather than shrubs. There are three species, the majority of which live in California's desert mountains. Leaves are all at the base of the plant and are fairly large, with three distinct main veins and the feel of puppy ears. Desert species have radiate heads with chaff that folds around the ovaries and falls with the fruit. Ray flowers are sterile and yellow. Disk flowers are numerous, yellow, and produce strongly compressed fruits with two small awns for a pappus.

Enceliopsis covillei (**Panamint daisy, sunray**) is rare, growing only on the west side of the Panamint Mountains at 400–

Enceliopsis covillei (**Panamint daisy**)

Enceliopsis nudicaulis (naked-stemmed daisy)

Ericameria cuneata var. *spathulata* (desert rock goldenbush)

1,250 m. We have seen it in stony canyons. It has large heads, up to 13 cm in diameter, with 20–35 ray flowers having ligules up to 5 cm. Leaves have winged petioles that merge into diamond-shaped or elliptic blades. *Enceliopsis covillei* is named for Dr. Frederick V. Coville (1867–1937), who led an important Death Valley expedition and frequently sent specimens to the National Herbarium.

Enceliopsis nudicaulis (**naked-stemmed daisy, sunray**) is "naked" (*nudi-*) "stemmed" (*caulis*), not unlike its cousin the Panamint daisy. It is uncommon, growing on stony hillsides and canyons of desert mountains at 950–2,000 m. Its heads generally have 21 ray flowers, with ligules up to 4 cm but usually closer to 2 or 3 cm. Petioles are barely or not winged, and leaf blades are oval.

Ericameria (**goldenbush**)—The genus name means "heath" (*erica*) "division" (*meri*), referring to its separation from a larger genus. The common name, goldenbush, describes the dense coating of yellow or golden flowers shrouding these dark green shrubs in either spring or fall. Heads may be radiate or discoid, and they have no chaff. Several species have heads with just a few ray flowers. Fruits are ribbed and have a pappus of simple bristles. Leaves usually are narrow, coated with resin, and sometimes also glandular. They frequently come in two sizes, either mixed along the stem or with larger leaves near the flowering branch tips. There are 27 species in western North America; 17 of these are found in California, of which eight occur in our deserts.

Ericameria cuneata var. *spathulata* (**desert rock goldenbush**) is named for the unusual shape of its leaves. Rather than being narrow, as is usual in *Ericameria,* its leaves are broad and "wedge-shaped" (*cuneata*), and in the variety *spathulata* the leaves are further rounded into a "spatula shape," often notched at the tip. Heads have 7–15 disk flowers, no ray flowers, and bloom in the fall. These plants grow in remarkable places, including narrow crevices of granite cliffs and other rock outcroppings at 100–1,900 m.

Ericameria linearifolia (**interior goldenbush, linear-leaved goldenbush**) has the most prominent ray flowers of desert species, and is spring blooming. There are 13–18 rays per head, with ligules 1–2 cm long. Plants are "linear" (*lineari-*) "leaved" (*folia*), with leaves 1–5 cm long—larger than most in the genus. Shrubs grow to 1.5 m in good conditions. They can be found up to about 2,000 m in the Mojave and western Sonoran Deserts.

Erigeron (fleabane daisy)—In addition to "woolly," as noted earlier, the Greek work *eri* can mean "early"; added to *geron*, meaning "old man," we have "aged early in life," as in being covered with white hair at an early age. The common name, fleabane—that is, keeping fleas away—is perhaps a more desirable characteristic. This large genus has almost 400 species around the world, 51 of which, including a few nonnatives, are established in California. All 10 of our desert species are natives. Most desert species are perennial, and most of them have radiate heads with white, pink, or blue ligules and no chaff. Fruits are generally two-ribbed, slightly hairy, and have a pappus consisting of 6–50 long inner bristles plus some shorter outer bristles or scales that are often difficult to detect.

Ericameria linearifolia (interior goldenbush)

Erigeron aphanactis (rayless daisy)

Erigeron aphanactis (rayless daisy, rayless fleabane) has a Greek name meaning that it does "not" (*a-*) "show" (*phan-*) its "rays" (*actis*). Discoid heads have many yellow disk flowers surrounded by many yellow pistillate flowers that have only a tiny or nonexistent ligule. There are just 12–17 pappus bristles. *Erigeron aphanactis* is a perennial from 8 to 20 cm tall, with leaves, stems, and phyllaries appearing fuzzy gray-green from a dense layer of stiff, spreading hairs. It grows in desert mountains from 1,300 to 2,600 m.

Erigeron concinnus (tidy fleabane) is a very different desert mountain species, growing mainly in rocky areas at 1,200–1,800 m. It has 40–60 ray flowers per head, with white, pink, or blue ligules that may be reflexed or coiled. The numerous disk

Erigeron concinnus (tidy fleabane)

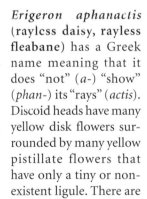

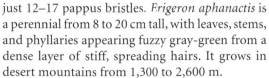

Eriophyllum pringlei (**Pringle eriophyllum**)

flowers are yellow. Fruits have a pappus of 7–15 bristles surrounded by a ring of obvious scales. Heads are generally dense atop short (6–16 cm) leafy stems. *Concinnus* means "tidy" or "elegant" and reflects the flat-topped, evenly spaced clusters of flower heads.

Eriophyllum (**woolly sunflower**)—The Greek meaning of the genus name, "woolly" (*erio-*) "leaf" (*phyllum*), is mirrored in the common name. The genus includes annuals to shrubs, and all 14 species are somewhat woolly. All are found in California, all are native, and six of them grow in our deserts. Ray flowers usually are yellow, but they may be white or absent. Disk flowers are yellow with a pappus of irregularly toothed or fringed scales, but in a few cases there is no pappus. Phyllaries are notable for being arranged in a single series, which means they stand side-to-side in one circle around the receptacle— sometimes partially fused. Each phyllary is paired with a ray flower.

Eriophyllum pringlei (**Pringle eriophyllum**) is named after an amazing botanist from Vermont, Cyrus G. Pringle (1838–1911), who made dozens of collecting trips to the desert southwest, including Mexico, sending specimens to Asa Gray for his work on *Synoptical Flora of North America.* For such a hardy and prolific collector, this is much too tiny a plant. It is an annual, spreading to only a few cm, and lacking ray flowers. Even the pappus is small (1 mm). It grows mainly in sandy areas among shrubs at elevations of 300–2200 m.

Eriophyllum wallacei (**Wallace eriophyllum, woolly daisy**) is named after William A. Wallace, a much less prolific collector than Pringle, who focused mainly on the area around Los Angeles in the 1850s. This attractive tufted annual can be as small as 1 cm or up to 15 cm tall, but it always has 5–10 ray flowers per head that glow like gold nuggets in low evening sun. There are many disk flowers. Pappus

Eriophyllum wallacei (**Wallace eriophyllum**)

is absent or less than 1 mm. Woolly daisy is common in our deserts under 2,400 m, usually in sand or loose gravel.

Geraea—There are just two species of *Geraea*, both in California, but only the more spectacular of them lives in the desert. The genus name means "old" and refers to the white-haired, aged appearance of the involucre or fruits.

Geraea canescens (**desert-sunflower**), our desert species, has radiate heads with 10–21 ligules about 2 cm long. Both ray and disk flowers are yellow, and receptacles bear chaff that is folded around flattened

Geraea canescens (desert-sunflower), Death Valley National Park near Beatty Junction

fruits. Fruits are hairy, up to 7 mm long, and have a pappus of two thin awns up to 4 mm long. Leaves and stems are covered with soft "graying" (*canescens*) hairs (sometimes stiff). The stems, which usually bear several heads, grow to about 50 cm. Plants are annuals, responding to good spring rains to make an impressive showing on sandy desert flats under 1,300 m that are bare in dry years.

Glyptopleura—This is a single-species genus with a name meaning "carved" (*glypto-*) "side" (*pleura*), for the curved and pitted surfaces of the fruit.

Glyptopleura marginata (keysia, crustleaf) is a beautiful and fragrant low-growing annual with ligulate heads of pale yellow flowers. Sap is milky, and leaves spread at ground level, displaying distinct white "margined" (*marginata*) lobes and teeth. There are

Glyptopleura marginata (keysia)

7–16 flowers per head with ligules about 1 cm long, and one to three heads per stem, although the stems are very short (1–5 cm). Keysia likes sandy flats of the Mojave Desert at elevations of 600–2,100 m. The common name is a tribute to hospitable early desert-edge residents named Keys.

Gutierrezia (snakeweed, matchweed)—The 25 species of *Gutierrezia*, named in honor of the Spanish noble family Gutierrez, are split between North and South America. There are three species in California, widespread in grasslands where they sometimes poison grazing livestock, though for the most part *Gutierrezia* is avoided by grazers and may be about the last thing left untouched. Two species extend into our deserts. They both are subshrubs, generally 30–60 cm tall and rounded out by numerous thin gummy branches and threadlike leaves. Both ray and disk flowers are few and yellow, but numerous heads may be scattered in small clusters over the branch tips. Receptacles have minute hairs but no chaff. Fruits have white hairs and a pappus of finely toothed scales.

Gutierrezia sarothrae (**broom snakeweed**) has at least two ray flowers per head, whereas its cousin *Gutierrezia microcephala* has just one or sometimes two. *Gutierrezia sarothrae* also has more (two to nine) disk flowers, as opposed to one to two in *Gutierrezia microcephala.* Otherwise the species are very similar. *Sarothrae* is from the Greek *sarothron,* "broom."

Helianthus (**sunflower**)—This is the genus named "sun" (*heli-*) "flower" (*anthus*), whose common name frequently represents the family as a whole.

Gutierrezia sarothrae (broom snakeweed)

There are 67 species of *Helianthus* in the Americas, 10 of which grow in California; three are nonnatives. One of the nonnatives and two natives are found in our deserts. Species may be annual or perennial, usually with an upright, fairly tall (about 1 m) stem, and broad three-veined leaves with stiff hairs. Heads are radiate and contain scaly chaff. Yellow ray flowers number 10 to many and are sterile. Disk flowers are fertile and numerous with corollas that vary from yellow to brown or purple. Fruits are flattened at the edges but plump where the seed is located, and the two pappus scales generally fall off as fruit matures.

Helianthus niveus ssp. *canescens* (**silver-leaved sunflower**) is the most representative of desert species and thrives in open sandy areas of the Sonoran Desert under 300 m. The other native species that

Helianthus canescens ssp. *canescens* (silver-leaved sunflower), Anza Borrego Desert State Park, Coyote Creek area

occurs in California's deserts is *Helianthus annuus*, with a range encompassing nearly the whole country and numerous cultivated forms. *Helianthus niveus* is named for its "silvery" or "snowy" (*niveus*) leaves covered with stiff white hairs. The endangered variety *tephrodes*, growing only on the Algodones Dunes, has long silky hairs.

Heliomeris (**golden-eye**)—The name means "sun" (*helio-*) "part" (*meris*), as in ray flowers being parts of the sun. There are six species in North America, with two in California. Only the California native is found in our deserts. Plants may be annual or perennial with slender stems and—unusual for the family—typically opposite leaves. Heads are radiate with chaff. Both ray and disk flowers are yellow-orange. Fruits have no hairs or pappus.

Heliomeris multiflora var. *nevadensis* (**Nevada golden-eye**) grows on rocky slopes and roadsides of desert mountains from 1,200 to 2,400 m. It has numerous stems, often reddish, with several flower heads per stem, so that an individual plant is "many" (*multi-*) "flowered" (*flora*). Leaves are long (2–6 cm), thin, sparse, and most are opposite. There are 8–20 ray flowers per head, with ligules to 2 cm long. The tips of ligules whiten quickly with age, giving a two-toned appearance to older flowers.

Heliomeris multiflora var. *nevadensis* (Nevada golden-eye)

Hulsea—All eight species of *Hulsea* grow in the southwestern United States, but only two occur in California's deserts. The genus is named after army surgeon and botanist G. W. Hulse (1807-1883), who discovered and collected one of its members. Heads have yellow to red ray flowers and yellow to orange disk flowers forming a dome. There generally are two pairs of pappus scales.

Hulsea vestita ssp. *inyoensis* (**Inyo hulsea**), considered rare in California, grows only in the eastern Sierra Nevada and northern desert mountains at 1,700–3,000 m, usually in the sandy bottoms of deep washes or canyons. Ray ligules are 12–18 mm long and yellow, matching the disk corollas. *Vestita* means "clothed" or "adorned," a reference to the dense long hairs that make leaves feel like velvet.

Hymenoclea (**winged ragweed**)— There are four species of *Hymenoclea*; two live in California, both in our deserts. The Greek name refers to a "membrane" (*hymen*) "enclosing" (*-clea*) the fruit and then flaring out like wings. Flowers on *Hymenoclea* are of a single sex and separated into single-sex heads. Pistillate heads have a single flower without a corolla and chaff that fuses around the fruit and widens above into winged

Hulsea vestita ssp. *inyoensis* (Inyo hulsea)

Hymenoclea salsola (burrobrush), fruits

membranes. Staminate heads have several flowers with tiny translucent corollas and free, rather than fused, anthers. Heads of a given sex tend to cluster together, but they generally share the same shrub or subshrub.

***Hymenoclea salsola* (burrobrush, cheese-bush)** is a very common inhabitant of both deserts under 1,800 m, especially in sandy locations. It is large (up to 2 m) for a subshrub and has many flexible, thin branches, green near the tips

Hymenoclea salsola (burrobrush), male flowers

and light tan near the base. Some people think it has an odor like stinky cheese. Although pistillate flowers have no corolla, the reddish tint of maturing membranous scales often attracts attention. Staminate flowers can be spotted due to their pollen and separate anthers. *Salsola* is the genus name for tumbleweed in the Goosefoot Family (Chenopodiaceae) and references the similarity of wings around fruits.

Hymenoxys (**goldflower**)—There are 28 species of *Hymenoxys*, all in the Americas, with four of them present in California and three of these growing in our deserts. The Greek genus name describes its "sharp" (*oxys*) "membrane" (*hymen*), which is the pappus of five scales with sharp pointed tips. Plants may be annuals to perennials, usually with radiate heads and no chaff. Both ray and disk flowers are yellow, and ray ligules are particularly wide with distinctive lobes at the tip.

***Hymenoxys acaulis* var. *arizonica* (Angelita daisy)** is the only species of *Hymenoxys* in California with basal leaves that are entire (not lobed or toothed) and with single flower heads on leafless

Hymenoxys acaulis var. *arizonica* (angelita daisy)

stalks that do not qualify as stems—hence the species epithet "without" (*a-*) "a stem" (*caulis*). Plants are perennial, branched only at ground level from a caudex that may be slightly woody. Narrow leaves, 2–6 cm long, are all at the base of the plant and are covered with soft hairs that tend to shed with age. Each head has many disk flowers and 8–13 ray flowers, with ligules up to 1.5 cm long. Even though the plants are no more than 20 cm tall with leaves just at the base, the flowers will catch your eye.

Isocoma (**goldenbush**)—This mixed Greek-Latin name means "equal" (*iso-*) "hair" or "hair-tuft" (*coma*), from having a pappus usually with two equal series of bristles twice as long as the fruit. There are about 10 species of *Isocoma* in southwestern North America; three are found in California, one of which inhabits our deserts. All are subshrubs with gland-dotted leaves and clusters of small heads of yellow disk flowers. Corollas, though tiny, are distinctively shaped with a very narrow tube at the base, expanding abruptly to a wider cylinder bearing erect lobes.

Isocoma acradenia (**alkali goldenbush**) likes fine sand or clay soils under 1,300 m that contain alkali or gypsum. Phyllaries of this species have a "pointed tip" (*acra*) that is "glanded" (*-denia*) and swollen like a wart. Plants are up to 1.3 m tall with several leafy

Isocoma acradenia var. *eremophila* (alkali goldenbush)

stems and branches having a shiny or polished yellowish look. Leaves of the variety *eremophila* extend all the way along branches to the flowering heads and have up to a dozen pointed teeth. Two other varieties have leaves without teeth or that are sparse near branch tips.

Iva (**wormwood**)—This genus with four species in California, two of which grow in our deserts, is named after a Mint Family (Lamiaceae) member, *Ajuga iva*, with a similar odor. Plants are annuals to shrubs with leaves opposite below but usually not so organized above. Heads contain two kinds of flowers plus chaff. Pistillate flowers are few to none and usually lack a corolla. Fruits lack a pappus. Staminate flowers are 5–20 with a translucent corolla and anthers that are free rather than following the family rule of being fused into a tube.

Iva axillaris ssp. *robustior* (**poverty weed**) is common in the central and western United States. It likes moist alkaline soils and is associated with poverty because of the poor ability of these soils to support agriculture and, therefore, those who

Iva axillaris ssp. *robustior* (poverty weed)

settled in those locations. Plants are perennial, spreading from shallow underground stems and growing erect to 10–50 cm. Linear to oval leaves grow all the way up stems, opposite near the base but alternate near the top, where each leaf axil has a single downward-hanging flower head. The specific epithet, *axillaris*, alludes to the flower position in the leaf axils. The subspecies name describes these plants as "strong" and "tough."

Lasthenia (**goldfields**)—Fields come alive each spring with the gold flowers of *Lasthenia*. A total of 17 species are found in North America and Chile, with 16 in California, four of which inhabit our deserts. Although the genus has both annuals and perennials, all our desert species are annuals with very narrow opposite leaves. Heads are radiate and without chaff. Pappus is awns, scales, or absent, often with variability within a single species. In desert species, corollas of both ray and disk flowers are yellow. One desert species, *Lasthenia microglossa*, has tiny or nonexistent ray flowers and four-

Lasthenia californica (goldfields), near the Antelope Valley California Poppy Preserve

lobed disk-flower corollas, making it unusual both for the genus and for the family. Lasthenia was the name of one of Plato's pupils.

Lasthenia californica (**goldfields**) is most likely to be seen carpeting large patches of the western Mojave Desert under 1,500 m. Beneath the deep yellow flowers, look for wiry reddish stems and linear opposite leaves. Plants can flower when they are only about 10 cm tall, but with good growing conditions they may get to 40 cm and be branched with numerous flower heads. Heads have 6–13 ray flowers with ligules up to 1 cm long, and many disk flowers. Pappus varies from none to narrow or wide awns.

Layia (**tidy-tips**)—This genus was named in honor of George Tradescant Lay, an English naturalist who sailed with Captain Frederick Beechey (for whom the Beechey ground squirrel is named) on a voyage of exploration from 1825 to 1828. There are 14 species of *Layia* in western North America, all occurring in California, but only two are found in our deserts. All species are annuals with narrow alternate leaves. Heads have chaff and are generally radiate, with ray ligules either white or yellow. Ray flowers have fruits without a pappus; disk flowers are yellow, and fruits generally have a pappus. The common name describes the three neat lobes on the ends of ray ligules.

Layia glandulosa (white tidy-tips)

Layia glandulosa (**white layia, white tidy-tips**) is common in open areas of sandy soil under 2,700 m. Stems are "glandular" (*glandulosa*) and 4–60 cm high, the size depending mainly on availability of moisture. Leaves are narrow and up to 10 cm long, sometimes with irregular lobes. There are 3–14 ray flowers, each fairly wide, up to 22 mm long, and with three large rounded lobes.

Most plants have white ray corollas, but in some areas they tend toward yellow. There are 17–105 disk flowers, and their fruits have a pappus of 10–15 flat awns with plumose bristles at the base.

Lepidospartum (scale-broom)—In Greek, *lepido-* means "scales" and *spartum* borrows from the genus name *Spartium* for Spanish-broom, a Pea Family (Fabaceae) plant that has the look of a broom. There are three species, two of which grow in California, including our deserts. Both are "broomlike" shrubs, meaning that if you cut off a few branches and tied them to a stick you would have a pretty good broom. Heads consist of just a few disk flowers having yellow corollas with long tubes and long lobes. Heads are clustered at branch tips, and numerous pappus bristles become conspicuous as fruits mature.

Lepidospartum squamatum (scale-broom) is a "scaly" (*squamatum*), spreading, rigid shrub with new growth woolly but soon developing a glossy dark green surface. Leaves come in two types: the young woolly ones are about 1 cm long and oval, but the older ones growing on flowering stems are scalelike. Leaves grade into scalelike bracts below the flower heads and then into similar phyllaries. Scale-broom likes washes and terraces under 1,800 m where there is a reliable source of groundwater.

Lepidospartum squamatum (scale-broom)

Lessingia (vinegar-weed)—A German specialist in Asteraceae, C. F. Lessing, is honored in this genus name. The genus, all 14 species of which occur in California, is extremely variable, including annuals to subshrubs, a wide variety of leaf forms, and heads that may or may not have ray flowers, with corollas that are white, yellow, or purple. It is therefore convenient that there is only one desert species.

Lessingia lemmonii (lemmonia) could have been named for the lemon color of its disk flowers, but John Gill Lemmon (1832–1908), an early forester who collected and wrote about southwestern plants, is the actual inspiration. *Lessingia lemmonii* is an annual, usually with thin spreading branches to about 20 or 30 cm, though they can be much smaller if conditions so dictate. The whole plant has an odor a bit like vinegar. Leaves along stems are narrow and only about 1 cm long, occasionally with a few teeth. Leaf tips are

Lessingia lemmonii (lemmonia)

sharply pointed, and when plants dry they become seriously prickly. Heads are one to a branch and have only disk flowers. Long narrow involucres are made up of numerous glandular phyllaries arranged like shingles but with their tips curved strongly outward. *Lessingia lemmonii* grows in sandy soil all over the Mojave Desert at elevations of 200–1,850 m. There are three intergrading varieties.

Machaeranthera (**hoary aster**)—*Machaer* is Greek for "sword" or "dagger" and describes the shape of the "anthers" (*anthera*) of this highly variable genus, the 35 species of which have almost nothing else in common. Seven of the eight species living in California are found in our deserts, so be prepared for variety, but this makes species keys easy to use. Getting to genus is not that hard either. Leaves tend to be narrow ovals with teeth or lobes that have sharp tips. Phyllaries are usually in several unequal series, arranged like shingles with straw-colored to purplish bases and green tips that spread outward or curl backward. There is a pappus of numerous bristles of differing lengths. Although there are exceptions to the above, after a while you will develop an eye for diagnostic features. Several species have ray flowers with pink to purple or bluish ligules, but two species have yellow ligules, one species has white to lavender, and another species has none.

Machaeranthera carnosa (shrubby alkali aster)

Machaeranthera carnosa (**shrubby alkali aster**) is the odd species with no ray flowers. It is a heavily branched subshrub growing to 50–90 cm, with green stems and scalelike leaves, hardly distinguishable from the stems. There are also some lower leaves that are up to 2 cm long, linear, and somewhat "fleshy" (*carnosa*). Heads grow one to a branch tip and contain 12–18 disk flowers that are deep yellow and sometimes tinged reddish. Fruits are almost cylindrical and have numerous ribs and hairs. The pappus is about 6 mm long. True to its common name, this shrubby aster likes alkali soils and grows in such parts of the Mojave Desert at 100–1,600 m.

Malacothrix—There are 21 species of *Malacothrix*; 14 of them live in California, with five occurring in our deserts. All have milky sap and ligulate heads without chaff but sometimes with thin fragile bristles—hence the Greek name "soft" (*malaco-*) "hair" (*thrix*). Ligules may be yellow, pale yellow, or white and often have lines or edges of red or purple. Fruits usually have a few stiff outer pappus bristles that remain attached plus 12–32 inner bristles that fall away as one. Leaves have no petioles, and sometimes their bases almost wrap around the stems. Larger leaves are basal, but smaller leaves usually extend upward to near flower heads. Leaves can be narrow and threadlike or broad, but they often have lobes and sometimes teeth. Plants may be annual or perennial. Desert species have most of the above variations.

Malacothrix coulteri (snake's head)

Malacothrix coulteri (**snake's-head**) has the look of a snake's head because of the large rounded involucres having phyllaries with dark central markings like eyes. The resemblance is not strong, but if you are nervous about snakes you will get the picture. This is a branched annual growing to 30 or 40 cm in good years. Stems and leaves are without hairs but usually have a whitish surface that comes off where touched. Leaves have broad lobed bases that wrap part way around the stems. Corollas are pale yellow and longer toward the outer edge of the head. *Malacothrix coulteri* likes sandy soils of the Mojave Desert under 1,500 m. It is named after Thomas Coulter (1793– 1843), an Irish botanist who explored in the Southwest in the 1830s; he was one of the first botanists to appreciate and describe the Colorado portion of the Sonoran Desert, encompassing the lower Colorado River basin.

Malacothrix glabrata (**desert dandelion**) is an annual you will see everywhere up to about 2,000 m. Practically a weed in good years, it looks a bit like huge clumps of dandelions—so pretty that you would be happy to have them in your garden. It grows to about 40 cm, with narrow, deeply lobed, "nearly hairless" (*glabrata*) leaves, mostly at the base, and numerous flower heads in full view. Each head is dense with yellow corollas.

Monoptilon (**desert-star**)—This genus of just two species is named for its pappus,

Malacothrix glabrata (**desert dandelion**)

which, in the species *Monoptilon bellidiforme,* looks like a "single" (*mono-*) "feather" (*ptilum*). It is really a single bristle with a plumose tip (plus a minute ring of scales), and it is a dead giveaway for this prostrate annual with white ray corollas growing in sandy washes of the Mojave Desert. The other species is superficially similar but has a more ordinary pappus. Desert-star describes white ray flowers covering the desert floor the way stars fill the sky.

Monoptilon bellioides (**desert-star, belly-flower**) is the more common of the two species. It is the original "belly" flower, because you have to get down on your

Monoptilon bellioides (desert star)

bellioide to see it up close. In reality, *bellis* is Latin for "daisy," and the ending *-oides* means that it "resembles" a daisy, which of course it does, with white ray corollas and yellow disk corollas. The pappus of *Monoptilon bellioides* consists of 0–12 bristles that are only 1–2 mm long but not all the same length, plus some even shorter scales that are dissected into bristles. *Monoptilon bellioides* sometimes carpets sandy ground in washes or other seasonally moist habitats from 200 to 1,200 m.

Nicolletia—Only one of the five species of *Nicolletia* grows in California. The whole genus is strongly and unpleasantly scented. Leaves have several narrow lobes, each with an embedded oil gland and a sharp bristle at the tip. Phyllaries have similar glands. The genus bears the name of French astronomer and geologist Joseph N. Nicollet (1786–1843), who was employed by the U.S. government to explore western territories.

Nicolletia occidentalis (hole-in-the-sand plant) is easily spotted even before you smell it. Its heads have 8–12 ray flowers that are deep rose-pink with lighter tips or bases. Golden disk flowers are almost 1 cm long and often tinged reddish. Just to be sure, check the pappus for five long pointed scales alternating with five clusters of shorter bristles. Plants are perennials but low growing from deep taproots. They have a habit of appearing in sandy depressions, even in roadways. Thus the common name describes their habitat in the Mojave Desert at 600–1,400 m, and *occidentalis* says they are "western."

Nicolletia occidentalis (hole-in-the-sand plant)

Palafoxia (**Spanish needle**)—Here is another genus making its home in the southwestern United States and Mexico. It is named for the Spanish General José Palafox (1780–1847). There is only one species in California.

Palafoxia arida var. *arida* (**Spanish needle**) is a branched annual growing to about 50 cm in low, sandy, "dry" (*arida*) parts of both deserts. Heads have 9–20 pink disk flowers, with corollas that are 1 cm long and have widely spreading, pointed lobes. Anthers are deeper pink to purple. Fruits are 10–15 mm long, and outer ones have just a few or no scales, while inner fruits have a pappus of spreading scales making dried fruits look

Palafoxia arida var. *arida* (Spanish needle) (left) and *Palafoxia arida* var. *gigantea* (giant Spanish needle) (right)

Pectis papposa var. *papposa* (chinch-weed)

Perityle emoryi (rock-daisy)

almost like hardened corollas. Plants are hairy gray-green, with sparse long narrow leaves and long thin phyllaries. A rare giant variety, *gigantea,* grows to 2 m, only in the sand dunes near Yuma.

Pectis—The Greek name, meaning "comb," comes from leaves that have long hairs lined up along the margins like teeth on a comb. *Pectis* is a genus with opposite leaves and also with strong odors, though not necessarily unpleasant in small doses. Heads are radiate with both disk and ray flowers having yellow corollas. Disk flowers are two-lipped. Phyllaries are in just one circle, and each ray flower has its own phyllary. Both leaves and phyllaries have embedded glands. There are 85 species, only one of which occurs in California.

Pectis papposa var. *papposa* (chinch-weed) is a low-growing, spreading annual that usually blooms over broad low patches of both deserts following summer rains. Leaves are dotted with glands along their margins, and phyllaries have a large gland near their tip and smaller marginal glands. The pappus (*papposa*) is a ring of about 20 bristles that are slightly plumose.

Perityle—There are 75 species of *Perityle,* and as with so many desert genera of Asteraceae, their home is exclusively in southwestern North America. Four species of *Perityle* occur in California, with three of these inhabiting the deserts. The genus includes annuals to shrubs, with leaves that are opposite or alternate, entire or deeply lobed to compound, and with or without petioles. Heads may be radiate or not, with either ray or disk flowers being yellow or white. So what do they all have in common? Disk flower corollas with four lobes is a common feature of *Perityle,* a character unusual within the family. The Greek name also describes a somewhat unusual fruit having a "surrounding" (*peri-*) "knob" or "callus" (*tyla*), meaning the swollen margin of a flattened fruit, sometimes with neatly aligned hairs. The pappus is

Perityle megalocephala (rayless rock-daisy)

a ring of scales, sometimes also including one or two bristles. Much of the variability within the genus is displayed by the few desert species that we have in California.

Perityle emoryi (rock-daisy) is the most common and easily recognizable member of the genus. It looks like a daisy, with white ray flowers and yellow disk flowers, and it grows in the rocks along canyon walls and on rocky slopes below 1,000 m in both deserts. It has broad, palmately lobed, strikingly green leaves that seem out of place in the desert, but it works because rock-daisy is an annual depending on abundant moisture of short duration. It generally looks best where it gets some afternoon shade from canyon walls. Although leaf form is somewhat variable, the palmate veins and lobes with irregular edges are distinctive. Look for them on fleshy branched stems that are both hairy and glandular. If you get too involved, you will notice a bad odor, and the brittle stems will break. The species name honors Major W. H. Emory (1812–1887), who directed the Mexican Boundary Survey.

Perityle megalocephala (rayless rock-daisy) is entirely different in its overall appearance from its cousin *Perityle emoryi*. Rayless rock-daisy is a subshrub, extensively branched into a rounded green mass up to 60 cm tall. Leaves are narrow, with a blade only 7–15 mm long, and rarely lobed or toothed. Heads of yellow disk flowers and no ray flowers grow in small clusters near branch tips, accounting for the specific epithet meaning "big" (*megalo-*) "headed" (*cephala*). Two varieties are recognized in our desert mountains, occurring at elevations of 1,300–2,800 m.

Peucephyllum (desert-fir, pygmy-cedar)—This is a single-species genus with a Greek name meaning that its "leaf" (*phyllum*) is similar to that of a "fir tree" (*peuce*).

Peucephyllum schottii (pygmy cedar)

Peucephyllum schottii (desert-fir, pygmy-cedar) is a large leafy dark green shrub up to 3 m in height but usually half that. Leaves are narrow and thick, like those of a fir tree, and shiny with resin and glands. Plants have an odor a bit like cedar, and sometimes the older wood curls and becomes sculpted like cedars. Heads grow singly on branch tips and are made up of 12–21 yellow disk flowers. Fruits are bristly and have a straw-colored to brownish pappus of numerous fine bristles. Pygmy-cedar likes to grow among rocks and boulders under 1,400 m. Arthur Schott (1814–1875) was a naturalist on the Mexican Boundary Survey.

Pleurocoronis (**arrow-leaf**)—Another small genus of southwestern North America, *Pleurocoronis* has three species, only one of which grows in California. Referring to the pappus, the Greek *pleuro-* (side) plus the Latin *coronis* (crown) means that the fruit has a crown of scales on one side. The common name describes the distinctively shaped leaves, with petioles as shafts and blades as points.

Pleurocoronis pluriseta (**arrow-leaf**) is the "side-crown" species with a "side-bristle" (*pluri-* plus *seta*). In addition to a crown of 10–12 scales, the pappus has 10–16 bristles. Heads have 25–30 disk flowers with pink to reddish corollas. Phyllaries are glandular with dark green tips that recurve. Plants are subshrubs about 40–50 cm tall, with numerous delicately curving green stems and odd leaves that take close inspection. Diamond-shaped blades are only 3–10 mm long and coarsely toothed, attached to a much longer, flattened petiole. Once you discover the leaves there can be no doubt about genus or species.

Pluchea—This is a largely tropical genus of 40 species named after the French naturalist N. A. Pluche (1688–1761). The genus has annuals to shrubs with disciform

Pleurocoronis pluriseta (arrow-leaf)

heads containing pistillate and bisexual flowers with pink or purple corollas. There are only two species in California.

Pluchea sericea (**arrow weed**) is a large shrub growing to several meters in dense clumps of stiff upright stems that were sometimes used as arrow shafts by local hunters. Oval leaves 1–4 cm long grow densely along each stem. Clusters of heads at stem tips contain flowers with pink corollas. Small, smooth fruits have a pappus of slender bristles. Plants grow only under 600 m in areas with water such as stream bottoms and springs. They have a high tolerance for salt and grow in dramatic patches around saline flats in Death Valley. *Sericea* is Latin for "silky."

Pluchea sericea (arrow weed) in Death Valley National Park, west of Salt Creek, below sea level

Porophyllum gracile (odora)

Porophyllum—This genus is characterized by its embedded oil glands and strong odor. "Pore" (*poro-*) "leaf" (*phyllum*) describes the dark oval glands that dot both leaves and phyllaries. Of about 30 species in North and South America, only one grows in California.

Porophyllum gracile (**odora**) is a subshrub with a strong, bad "odora." It is usually 40–50 cm tall, with numerous "thin" or "slender" (*gracile*) stems and leaves. Five phyllaries surround 20–30 disk flowers with whitish to purplish corollas. Fruits are almost 1 cm long, with a somewhat shorter pappus of many bristles. Odora grows on rocky slopes and in small washes under 1,500 m in both deserts.

Psathyrotes—This genus has five species in southwestern North America, two of which inhabit California. They are all very low growing "rounded" (*-rotes*) shrubs with numerous "brittle" or "crumbling" (*psathy-*) branches, and they have a scent like turpentine.

Psathyrotes ramosissima (**turtleback**) has broad, hairy gray-green leaves that form a rounded mound like a turtle's back, with leaves for plates—hence its common name. *Ramosissima* is Latin meaning that this is the "most branched" of the species. It also has more flowers per head (21–26) and more pappus bristles (over 120) than the other California species, *Psathyrotes annua* (fan-leaf), of the Mojave Desert. In both species, heads consist only of yellow disk flowers with no chaff. *Psathyrotes ramosissima* is common in sandy areas under 1,000 m in both deserts.

Psilostrophe (**paper-daisy**)—Here is another small genus of southwestern North America. There are six species, only one of which grows in California. The genus is characterized by radiate heads with two to six ray flowers having deep yellow corollas that fade to a papery texture and hang downward as they dry. Ligules are broad, with three lobes at their tips. Disk flowers are yellow and hairy, and fruits have a pappus of four to six transparent scales. The Greek name means "to turn" (*-strophe*) "bare" or "naked" (*psilo-*).

Psilostrophe cooperi (**paper-flower, Cooper paper-daisy**) is a perennial to subshrub to about 60 cm, with numerous white-hairy branches and linear leaves 1–8 cm long. Heads have 3–6 ray flowers and 10–25 disk flowers. It grows in the eastern Mojave and Sonoran Deserts at 150–1,500 m. Dr. J. G.

Psathyrotes ramosissima (turtleback)

Psilostrophe cooperi (paper-daisy)

Cooper participated in the Geological Survey of California and collected plants from the Mojave Desert in 1861.

Rafinesquia (**chicory**)—There are two species of *Rafinesquia,* and both grow in California's deserts. Plants are annuals with milky sap and ligulate heads with no chaff. Corollas are white, some with purplish lines on the underside. Deeply lobed leaves grow up stems but are larger and more lobed near the base of the plant. Lobes are pointed but not spiny. Fruits taper to a beak that terminates in a ring of plumose bristles. Constantine S. Rafinesque (1783–1840) has been described as an eccentric American naturalist and scholarly recluse.

Rafinesquia neomexicana (**desert chicory**) has ligules 15–20 mm long, about 10 mm longer than the phyllaries. *Rafinesquia californica* has ligules only 5–8 mm long, just 3–5 mm longer than the phyllaries. Both species grow in both deserts under about 1,500 m. Look for them spreading up through shrubs to support their stems.

Senecio (**groundsel, ragwort, butterweed**)—With roughly 1,500 species, this probably is the largest

Rafinesquia neomexicana (desert chicory)

genus of Asteraceae, and it grows all over the world. California has 42 species, several of which are nonnatives. Our deserts have five, all of which are native. The most distinctive feature of the genus as a whole is the single circle of upright, equally sized phyllaries surrounded (usually) by a few much smaller and poorly organized phyllaries that spread away from the stem. Heads may be radiate or discoid without chaff, and corollas are almost always yellow to orange. Fruits are cylindrical, frequently hairy, and topped by a pappus of minutely barbed bristles that fall readily. Leaves are alternate and generally transition from large with petioles at the base of stems, to bractlike near the top. Plants can be annuals to trees, but in our deserts we have annuals to subshrubs. The Latin word *senex,* meaning "old man," provides the root for the genus name and refers to the whitish beardlike pappus.

Senecio flaccidus var. *monoensis* (**sand-wash groundsel**) is a subshrub with upper branches that arch over from their own weight, providing the specific epithet *flaccidus* (limp). Plants usually grow to about 1 m tall. Leaves are threadlike and divided all the way down to the veins. They, too, are "flaccid." Clusters of radiate heads, usually with 8 ray flowers and fewer than 40 disk flowers, bloom persistently at almost any time of year but mainly in spring. Sand-wash groundsel is common in sandy washes of both deserts at elevations of 500–2,000 m.

Solidago (**goldenrod**)—There are about 150 species of *Solidago,* concentrated in North America. California has 10, of which only one grows in the deserts. They are all perennials with upright stems branched near the top. Leaves, diminishing in size upward, are usually oval and resinous. Small flower heads grow along several short, clustered branches at the tops of stems. Usually heads are concentrated just on one side of the small branches. Flowering branch tips are the "golden" part of the "rod." If there are ray flowers, their ligules are very short and inconspicuous. Both ray and disk flowers are yellow. Fruits have a pappus of bristles with long barbs. *Solidago* means "to make well or

Senecio flaccidus var. *monoensis* (sand-wash groundsel)

Solidago confinis (southern goldenrod)

whole" and refers to medicinal qualities known to early cultures.

Solidago confinis (**southern goldenrod**) can be 2 m tall, with lower leaves up to 25 cm in length. Each head has 10–20 disk flowers and 3–13 ray flowers with ligules under 3 mm. Southern goldenrod likes marshy spots and wet stream banks under 2,500 m in the northern desert mountains. *Confinis* is a Latin word meaning "bordering on" or "adjoining," possibly a reference to its distribution.

Sphaeromeria—This is a genus of nine species from western North America, two of which grow in California. Our one desert species slides in from the north, liking high elevations among rocks. Plants are perennials or subshrubs with resinous glands but made silky by long hairs attached by a short off-center trunk. Heads have a few pistillate flowers surrounding more numerous disk flowers. There is no chaff or pappus (usually). The Greek name, meaning "spherical" (*sphaero-*) "part" or "division" (*meri*), indicates that this genus was separated from a larger group of related species (previously classified as *Tanecetum*).

Sphaeromeria cana (**alpine tansy**) is a subshrub generally 10–20 cm high with three to eight heads in dense clusters at the tips of the tallest branches. These stems have narrow oval leaves about 1 cm long, while leaves closer to the base of the plant have a few lobes. Fruits are 10-ribbed and without any pappus. Dense hairs give this species its *cana*, or "ashen-gray," coloring. Alpine tansy is uncommon in the northern desert mountains at elevations of 3,000–4,000 m.

Stenotus—This is a genus of mat-forming perennials, usually growing at mid- to high altitude, and generally with leaves concentrated near branch tips. There are five species, all in western North America, with three in California, one of which is found in our desert mountains. Heads are radiate, with both ray and disk corollas yellow. Fruits are generally hairy and tipped with a pappus of soft bristles. The Greek name means "thin" or "narrow" (*steno-*) "ear" (*otus*).

Stenotus acaulis (**short-stemmed stenotus**) has narrow three-veined leaves up to 10 cm long. There are 6–15 ray flowers with ligules to 12 mm, and 25–50 disk flowers in each head. Fruits come with or without long silky hairs. *Acaulis* means "without a stem," but there are plenty of short stems and branches.

Sphaeromeria cana (alpine tansy)

Stenotus acaulis (**short-stemmed stenotus**)

Stephanomeria paucifl..ra (wire-lettuce)

Stephanomeria (wire-lettuce)—This genus has 24 species, all in western North America; 12 are in California, and four of these in our deserts. All species have milky sap and ligulate heads with pink or lavender to almost white corollas. The common name refers to the generally wiry stems of this relative of lettuce. The Greek name describes this genus as the "wreath" (*stephano-*) "division" (*meri*), segregated from a larger group.

Stephanomeria pauciflora (wire-lettuce, desert-straw) is the species "poor" (*pauci-*) in "flowers" (*flora*). There are only five or six (sometimes fewer) flowers per head, and they can be deceptive because the five ligules look very much like the five petals of a single flower. Plants are perennials to subshrubs, around 50 cm tall and rounded

Stephanomeria pauciflora (wire-lettuce)

with densely branched rigid straw-colored (when dry) stems, tasty to donkeys. Leaves are totally inconspicuous. Desert-straw is common under 2,400 m.

Stephanomeria spinosa (spiny wire-lettuce) is a subshrub under 40 cm in height, with networks of stout branches that terminate in sharp points or "thorns" (*spinosa*). Similar to *Stephanomeria pauciflora*, it has only three to five flowers per head, but its weaponized branches make it easy to identify. It likes elevations of 1,200–3,300 m in our desert mountains.

Stephanomeria spinosa (spiny wire-lettuce)

Syntrichopappus (xerasid)—There are only two species of *Syntrichopappus,* and they both live in California; just one is a desert species. The long Greek name describes the "joined together" (*syn-*) "hair" (*tricho-*) of the "pappus," meaning many bristles fused at the base. However, some species have no pappus. *Xerasid* is also Greek, somewhat abbreviated, meaning "son of dryness."

Syntrichopappus fremontii (Fremont xerasid) is named after our intrepid explorer John C. Frémont (1813–1890), whose name is honored by many species as well as a genus. Fremont xerasid is an annual with spreading stems no more than 10 cm tall. Its leaves are woolly hairy and spoon-

Syntrichopappus fremontii (Fremont xerasid)

shaped, arranged opposite near the base of stems and alternate above. Heads are radiate with one ray flower per phyllary. Ligules are three-lobed and yellow. The fused pappus has 30–40 bristles. Look for this species in sand or gravel of the Mojave Desert from 600 to 2500 m.

Tetradymia (**cotton-thorn, horsebrush**)— There are 10 species of *Tetradymia* in western North America, eight of which are in California, and five in our deserts. Cotton-thorn is a good description for several of these shrubs famous for their spines and covered with dried cottony pappus on mature fruits. Spines form from persistent leaf veins after the blades fall away, so the youngest stems and several species have no spines. Pappus also is lacking in some species, although our four desert dwellers have a good pappus of fine bristles, and the fifth has scales. "Four" (*tetra-*) "together" (*-dymia*) aptly describes most flower heads in the family—usually a neat package of four cream to yellow disk flowers with long, spreading corolla lobes.

Tetradymia glabrata (bald-leaved cotton-thorn)

Tetradymia glabrata (**bald-leaved cotton-thorn**) is one of two desert species that has no thorns (spines). It grows to about 1 m and has narrow leaves 1 cm long, growing individually along new stems and flowering stems. Other leaves are clustered and threadlike. Both stems and leaves have woolly hairs when young, but then become "nearly hairless" (*glabrata*). *Tetradymia glabrata* grows in the Mojave Desert at 800–2,400 m, often beside dry sandy washes.

Thymophylla (**dyssodia**)—This genus has about 17 species, favoring southern North America and the Caribbean. There is just one species in California, found in the eastern Mojave Desert. The Greek name means "thyme" (*thymo-*) "leaves" (*phylla*), probably a reference to the strong odor conferred by embedded oil glands. Like *Adenophyllum*, *Thymophylla* was once in the genus *Dyssodia*.

Thymophylla pentachaeta var. *belenidium* (**Thurber dyssodia**) is a subshrub growing to 30 cm with numerous thin stems and opposite leaves. Leaves look like clustered needles, but each leaf

Thymophylla pentachaeta var. *belenidium* (Thurber dyssodia)

Trichoptilium incisum (yellow-head)

actually has several stiff, deeply divided lobes dotted with tiny glands. Often the upper stems and leaf tips are reddish or brown, looking dry and tired. Heads are radiate and generally have 13 ray flowers with yellow ligules less than 4 mm long. The many disk flowers also are yellow, and fruits have a pappus of 10 scales, each of which divides into three awns. Sometimes five scales are vestigial, leaving "five" (*penta-*) with "bristles" (*-chaeta*).

Trichoptilium (yellow-head)—This is a single-species genus named for it pappus, described as a "hair" (*tricho-*) that is "feathery" (*-ptilium*). This odd pappus consists of five elongated scales, each with a fringe of bristles, longest at the tip of the scale.

Trichoptilium incisum (yellow-head) is a very pretty annual that looks perennial and sometimes is. It has several stems spreading from the base and clustered leaves with a dense coating of curly hairs. Leaves are "incised" or "cut into" (*incisum*), making large teeth or small pointed lobes. About 10 cm

Trixis californica var. *californica* (trixis)

past the leafy part of the stems is a solitary hemispherical head of deep yellow disk flowers. Perimeter flowers are slightly bilateral and have enlarged corolla lobes on their outside edges. Yellow-head grows under 1,000 m in the southern Mojave and Sonoran Deserts.

Trixis—The genus of tricks, *Trixis* is odd for having a head of two-lipped disk flowers with one of the lips sufficiently elongated to be ligule-like. Thus the heads, technically discoid, appear radiate or ligulate. The Greek word *trix* means "threefold," referring to the lobes of the elongated corolla lip. There are about 65 species in North and South America, but only one grows in California.

Trixis californica var. *californica* (trixis) is the species "of California." It is a shrub without milky sap—features that distinguish it from all California Asteraceae members with ligulate heads, which *Trixis californica* otherwise resembles. Heads have

11–25 two-lipped yellow disk flowers. Each corolla is tubular for 6–9 mm at the base and then splits into two lips, the longest of which is 5–8 mm, or the equivalent of many ray flower or ligulate flower ligules. Shrubs are rounded, leafy, and glandular, usually around 1 m tall. Leaves are dark green narrow ovals, 2–11 cm long, either with or without teeth even on the same plant. Trixis grows under 1,000 m, often in sandy washes or in partially shaded rocky margins of washes.

Viguiera—There are around 150 species of *Viguiera*, all in the Western Hemisphere, but only three in California, of which two inhabit our deserts. The name honors a French physician and botanist, Dr. L. G. A. Viguier (1790–1867), who lived in Montpellier, home of one of Europe's oldest botanical gardens, which was part of the medical school. *Viguiera* may be annuals to shrubs with radiate or discoid heads that have chaff. Fruits are flattened and have a pappus of scales, with some usually long and pointed.

Viguiera reticulata (**leather-leaved viguiera**) has large leathery "reticulated" (*reticulata*) leaves, meaning that the underside has a network of raised veins sufficient to excite any phlebotomist. There are three main veins, and leaves are widely oval up to 9 cm long— much larger than those of Parish viguiera (*Viguiera parishii*), the other desert species. Both species are shrubs with radiate heads of

Viguiera reticulata var. *reticulata* (leather-leaved viguiera)

many yellow disk flowers plus 7–15 yellow ray flowers with ligules over 1 cm long. Heads are solitary or in small loose clusters at the tips of long leafless stems. Both species have a pappus of two short scales plus two (or three) longer pointed ones. *Viguiera reticulata* grows in the Mojave Desert under 1,500 m, often at the base of canyon walls.

Xylorhiza (**desert-aster**)—All eight species of *Xylorhiza* (a genus name meaning "woody" [*xylo-*] "root" [*rhiza*]) grow in Western North America, with three of them in California and our deserts. Two of our species are rare and of extremely limited distribution. The genus as a whole and our California representatives cover the range of perennials, subshrubs, and shrubs, which is one of the main ways of distinguishing among them. Heads are radiate with many yellow disk flowers and 20 or more ray flowers with light blue to lavender ligules up to about 3 cm long. These large showy heads are raised above the leafy plants on their own bare stems, making an impressive statement in the landscape.

Xylorhiza tortifolia var. *tortifolia* (**Mojave aster**) is widespread in the Mojave and northern Sonoran Deserts, where it can be either a perennial or a subshrub. Leaves are linear to oval, covered with soft hairs, glandular, and either with or without spine-tipped teeth. Also, "leaves" (*folia*) are *torti-*, or "twisted," rather than flat. There are 25–60 ray flowers and even more disk flowers. Fruits are hairy and have a pappus of many unequal bristles up to 9 mm long. Plants grow at 240–2,000 m, frequently scattered across stony slopes and flats. Our rare species, *Xylorhiza cognata* and *Xylorhiza orcuttii*, are shrubs growing in their own special canyons at lower elevations than *Xylorhiza tortifolia*.

Xylorrhiza tortifolia var. *tortifolia* (Mojave aster)

Examples and Uses

In general, Asteraceae are not very edible, but you might not even realize that a few familiar foods are part of the family. Lettuce (*Lactuca*) is a good example. It is related to desert species having milky sap, a feature we do not see much in grocery store varieties. Milky sap is more noticeable if you pick lettuce fresh from your garden or if you let plants form flower stalks when the weather gets warm. Chicory and endive (*Cichorium*) are also in the milky-sap group.

The group that we think of as thistles provides artichokes (*Cynara*) and safflower oil (*Carthamus*). Artichokes are flower heads in the bud stage, the immature flowers being the "choke," well protected inside the bracts and phyllaries from which we scrape tasty morsels with our teeth. The "heart" is the receptacle. Sunflowers (*Helianthus*) provide both Jerusalem artichokes, a root expansion, and sunflower seeds, which are the little gray ovals we eat after the "shell," or dried ovary wall, has been removed. Finally, chamomile (*Anthemis nobilis*) tea is favored in some circles as an alternative to teas with caffeine, and may also have medicinal qualities. It is usually made from dried flower parts.

Several members of the family have provided treatments for various ailments, such as two species of *Artemisia* that can be used to treat intestinal parasites. *Baccharis* leaves contain antibacterial compounds and have been used medicinally by Indian populations to treat wounds and reduce swelling. Folk medicines are fairly common in the family, including a topical treatment for snake bites and teas for malaria from *Gutierrezia*. The organic gardener's insecticide pyrethrum comes from *Chrysanthemum cinerariifolium*.

Both delightful flowers and devilish weeds are included in Asteraceae, with *Zinnia* and *Dahlia* in the first category and dandelions (*Taraxacum*) and sow thistles (*Sonchus*) in the second. Many invasive thistles of various genera spoil heavily used grazing land and other disturbed areas. Their removal in habitat restoration projects is a daunting task.

Chilopsis linearis ssp. *arcuata* (desert willow), north side of the Clark Mountains

BIGNONIACEAE

BIGNONIA OR CATALPA FAMILY

The Bignonia Family includes some of the most beautiful flowering trees and vines grown in southern California. These include the *Jacaranda*, a common street tree with large blue to purple flower clusters in summer, and the trumpet creeper (*Campsis*), which climbs over fences and up telephone poles to show off its orange-red trumpet-shaped flowers. The only native member of the family in California is desert-willow, equally spectacular when covered with pink blossoms but little used in landscaping.

Abbé Jean Paul Bignon (1662–1743) is honored by this family name. He was librarian to King Louis XIV of France.

Bignoniaceae Icon Features

- Fruit from a superior ovary, elongated similar to a bean
- Fruit splitting lengthwise to release seeds with membranous wings
- Leaves long and thin, similar to the fruits

Distinctive Features of Bignoniaceae

Bignoniaceae species often have large flowers and beanlike fruits with winged seeds. Their five petals are fused into a tube with lobes either in radial symmetry or consolidated into two unequal lips. Stamens are attached to the inside of the corolla tube; although there are usually just four, sometimes a fifth, infertile one, known as a **staminode**, is present as a smaller filament without an anther.

There is a long style and the stigma has two flat, flaplike lobes designed to promote cross-fertilization (fertilization of ovules with pollen from a different flower). When a bee arrives, it is attracted to nectar near the base of the corolla tube. First, however, it encounters the spread-apart stigma lobes, and as it brushes by it deposits pollen from a different flower on the receptive inner surface. At this point the stigma lobes, which are sensitive to touch, fold together like closing a book, so that only the outer, nonreceptive surface remains exposed. The bee is then dusted with new pollen as it forages for nectar within the flower, and when the bee leaves, the closed stigma lobes prevent the flower's own pollen from being deposited on its stigma.

The ovary is superior and generally has two chambers and numerous ovules. Fruits are long, thin cylinders like beans, but they are not beans (legumes) mainly because the seeds are attached along both edges of two chambers instead of along just one side of a single chamber (see Fabaceae) as is the case in legumes. When the Bignoniaceae fruit dries, it splits open along opposite edges to release its seeds, which are often winged to assist in dispersal.

Bignoniaceae species usually have big leaves as well as big flowers, although often the large leaves are compound and divided into numerous tiny leaflets, like those of *Jacaranda*. Desert-willow (*Chilopsis linearis*), however, has a long, narrow leaf that, like a willow, is not compound. Most members of the family are woody, growing as trees or vines.

Chilopsis linearis ssp. *arcuata* (desert willow)

Similar Families

Some flowers in the Figwort Family (Scrophulariaceae) are quite similar to flowers of Bignoniaceae, however you will never have occasion for confusion in the desert: desert-willow is a tree, and there are no desert trees in the Figwort Family.

The superficial similarity of the Bignoniaceae fruit to a bean could have you considering Fabaceae as a family, but you will never find winged seeds inside a real bean. Flowers of desert-willow are entirely unlike any Fabaceae flowers, which have 5 separate petals (or a tiny tubular corolla) instead of the large tubular corolla of Bignoniaceae. Furthermore, California desert Fabaceae members that are trees or large shrubs have compound leaves or short simple leaves rather than long, simple leaves, and their flowers are yellow or lavender to purple rather than pink.

You might think desert-willow is in the Willow Family (Salicaceae), but only if you see it without its flowers or fruits. The big, pinkish, tubular flowers of desert-willow in no way resemble the small catkins, or pussy willows, of a real willow tree. Similarly, Salicaceae species never have long, bean-shaped fruits.

There is nothing else like it. If you see a tree that has big pinkish flowers and bean-shaped fruits but otherwise looks like a willow, it is desert-willow.

Family Size and Distribution

There are 110 genera and 800 species of Bignoniaceae, mainly in the South American tropics, although some are native to other tropical and a few temperate parts of the world. Some of the more decorative genera, such as the blue-flowered *Jacaranda,* have been spread throughout mild climates in landscaping. A few genera, including *Catalpa*, with frost-tolerant members, have a temperate-zone distribution, but they are the exception.

Chilopsis linearis ssp. *arcuata* is the only representative of Bignoniaceae native to California, and it is restricted almost entirely to the desert.

California Desert Genus and Species

As the name implies, desert-willow is found only where there are sources of water, such as at sites of seasonal concentration that provide a fairly consistent underground supply or along streams or springs with surface water. During dry periods it loses its leaves and may be hard to spot, but as water arrives it flowers in great profusion, with leaves following.

Chilopsis (**desert-willow**)—This single-species genus occurs only in low desert of the southwestern United States and northern Mexico. The Greek name means "resembling" (*-opsis*) a "lip" (*chil-*), an allusion to the two-lipped corolla.

Chilopsis linearis ssp. *arcuata* (**desert-willow**) is named for its long, narrow, "linear" (*linearis*) leaves, which are unusual in the family.

Examples and Uses

Yellow trumpet flower (*Tecoma stans*), growing in Mexico, Texas, and Arizona, has been used by locals to treat syphilis and diabetes, but its efficacy is unknown. Other uses of the Bignonia Family are mainly as described above for landscaping.

Cryptantha confertiflora (yellow cryptantha), north end of Death Valley National Park, Last Chance Range

BORAGINACEAE

BORAGE FAMILY

Forget-me-nots are the classic representative of the Borage Family. These charming small blue flowers are carried by children to their mothers and by lovers to each other with a universal message of bonding. You will not find forget-me-nots (*Myosotis*) in the desert, but you will not forget one of the largest, most common and confusing genera of the family, the dreaded *Cryptantha*. These small white flowers require a hand lens and a lot of patient practice to identify to species. But never mind about species! As a whole, the family is an important part of spring blooms and you will enjoy getting to know its members.

The family name follows the genus name *Borago*, which is simply the Latin word for these kinds of plants. The origin of *Borago* may be *borra*, meaning "rough hair"—a good descriptor for many members of the family.

Boraginaceae Icon Features

- Flowering stem tips coiled with flowers maturing sequentially as the coil unwinds
- Corollas five-lobed
- Plants often with stiff hairs

Distinctive Features of Boraginaceae

Borage Family members can usually be spotted by their coiled spikes of small flowers, maturing first at the base of the coil and progressing toward the tip as it continuously unwinds. However, this feature is not found in all Boraginaceae species—and there is another common family in the desert with the same habit, so read on. Fruits of most Boraginaceae members are distinctive, but not unique, for having four **nutlets**, each containing a single seed. The teardrop-shaped nutlets generally sit symmetrically around the flower style, which is usually attached at the base of the nutlets. Nutlets retain a diagnostic scar when removed. Features of the attachment scar and nutlet wall are important in distinguishing some genera and many species. In these cases, it is essential to have mature fruits as well as flowers for proper identification. Look for fruits low on the flower spike, and pay special attention to any prickles or appendages they may have.

Boraginaceae flowers are radially symmetrical with a tubular corolla that spreads at various angles and has five lobes. Usually there are five petal appendages at the point where the corolla tube spreads open. These form a crown that sometimes obscures the opening of the tube and often is a different color from the remainder of the corolla. It might look like the ring around the bull's-eye of a target. There are also five sepals, which may be either separate or fused only near the base. Calyx lobes generally elongate as fruits mature, frequently exceeding the length of the fruit, and sometimes providing a protective casing. The five stamens are attached to the corolla between lobes. Almost all Boraginaceae genera have a single style, or sometimes none at all, and a single stigma (although it may be obscurely lobed). Only the genus *Tiquilia* has a deeply divided style and two stigmas.

Boraginaceae species may be annuals, perennials, shrubs, or even trees, although shrubs and trees are not found in our deserts. Many desert species are annuals, but a few species live two or several years, developing a little woody tissue, sufficient in rare cases to qualify them as subshrubs. Stem and leaf hairs are commonly stiff and bristlelike, making the plants unpleasant to handle. Leaves may grow mostly at the base, but there usually are many leaves arranged alternately along the stems as well. Leaves are generally entire (without lobes or teeth) and a lot longer than wide, although this is not universally true, even in the desert.

Similar Families

The Waterleaf Family (Hydrophyllaceae) has numerous desert members with small flowers on one side of a coiled stem tip, just like Borage Family flowers. Plants may look similar as well, but two big differences will help you tell them apart. First, Hydrophyllaceae desert genera generally have two distinct style lobes or two styles rather than the single unlobed style of all desert Boraginaceae members other than *Tiquilia*. Second, fruits of Hydrophyllaceae do not consist of four nutlets, but are single spherical or oval capsules with several to many seeds.

Mint Family (Lamiaceae) fruits resemble the four-nutlet fruits of the Borage Family, but you will immediately see other features of mints that are very different from borages. Mint flowers generally grow in tight clusters, like balls, spaced along flower stems. There is nothing resembling a coil of flowers. Flowers of most Lamiaceae species have bilateral symmetry and would not be confused with the radial symmetry of Boraginaceae. Most mint stems are square in cross section, and leaves are opposite and highly aromatic. These features will not be found in borages.

A few desert *Verbena* species (Vervain Family, Verbenaceae) have four nutlets, but like the mints, they have opposite leaves and stems square in cross section—features that would be hard to confuse with Boraginaceae.

Family Size and Distribution

The 100 genera of Boraginaceae, comprising some 2,000 species, are particularly abundant in western North America and the Mediterranean. They like both temperate and tropical climates, and some genera occur throughout the Northern Hemisphere. *Cryptantha*, one of the largest genera, with 160 species, is well represented in California, and in our deserts in particular. *Heliotropium* is probably the largest genus, with about 250 species, and is widely cultivated as well as having a broad native distribution.

California has 18 genera of Boraginaceae, five of which consist solely of nonnative species. Desert representatives are found in nine genera, and all of our desert species are natives.

Boraginaceae members in California's deserts are overwhelmingly annuals. Despite having only four seeds per flower, plants are widespread and enormously abundant when there is enough rainfall for good germination and growth. In dry years good crops of flowers may be found in washes or roadside ditches, where a little runoff contributes to soil moisture.

California Desert Genera and Species

Almost all desert Boraginaceae species in California have rather small white or yellow flowers, and the coiled spikes, though not universal across the family, are easy to spot. Plants are usually under half a meter in height, and some are very much smaller than that, requiring a careful search. Even the perennials and subshrub are of low stature. Many plants are leafy, utilizing the sun's energy for the short time that water is available in the upper layers of soil. Annuals need to make seeds in a hurry because it is only through seeds that species survive long dry seasons.

SIMPLIFIED KEY TO DESERT GENERA OF BORAGINACEAE

Both flowers and fruits are necessary for many species identifications, but genus often can be determined based on flower features alone. Nevertheless, do not pass up an opportunity to look at mature fruits, which can often be found in combination with flowers because of the sequential maturation of flowers along a coiled spike. The oldest flower on a stem, and therefore the most likely source for a mature fruit, is at the joint where two or three coiled spikes originate.

1. Corollas pink, lavender, or purple, with a yellow tube when young. Style deeply divided into 2 branches, each with a stigma (4 spp) .... *Tiquilia*
1' Corollas white, cream, yellow, or orange. Style absent or undivided, stigma single.
  2. Corollas yellow, sometimes orange or with orange markings.
    3. Plants annual. Corolla deep yellow to orange or with orange markings (3 spp) ...... *Amsinckia*
    3' Plants perennial. Corolla yellow to lemon yellow.
      • Corolla tube 15–35 mm, lobes jagged-margined, not overlapping or touching, yellow ....
      .... *Lithospermum incisum*
      • Corolla tube 9–13 mm, lobes smooth-margined, overlapping or at least touching edges, lemon yellow or light yellow .... *Cryptantha confertiflora*
  2' Corollas white or cream, appendages in central ring sometimes yellow.
    4. Sepals maturing differently, with upper 2 partly fused, greatly enlarging, and growing 5–10 sturdy spines with sharp hooks along their edges. Nutlets 2, maturing differently, protected by burlike fused sepal pair .... *Harpagonella palmeri*
    4' Sepals maturing similarly, some slightly larger but not spectacularly different. Nutlets 4, or initially 4 with some not maturing but still visible.
      5. Style attached to top of ovary, falling away as nutlets mature (4 spp) .... *Tiquilia*
      5' Style attached at base of ovary, generally visible at center of the 4 nutlets.
        6. Nutlet margins fringed with long pointed or hooked appendages.
          • Nutlets widely spreading into an X-shape, often irregular; marginal appendages pointed or curved, but not with barbs at the tip (7 spp) .... *Pectocarya*
          • Nutlets mostly erect, not widely spreading, marginal appendages stout spikes with barbed tips .... *Lappula redowskii*
        6' Nutlet margins rounded or angled (rarely fringed and, if so, fringe mainly entire and not divided into pointed appendages).
          • Nutlet attachment scar generally recessed, like a groove; nutlet wall above the attachment scar also grooved and slightly spread open, forking at the base (26 spp) ....
          .... *Cryptantha*
          • Nutlet attachment scar generally elevated, like a ridge (keeled); nutlet wall above attachment scar also keeled (10 spp) .... *Plagiobothrys*

Amsinckia (fiddleneck)—This genus of 10 species is named after Wilhelm Amsinck, a 19th-century patron of the Hamburg Botanic Garden. Species are restricted to western portions of North and South America, with nine of them found in California. Three of these are represented in our deserts. Fiddleneck is an excellent common name, since the coiled flower spike looks much like a fiddle's neck. *Amsinckia* are always annuals, generally (but not always) with stiff hairs all over the green stems, leaves, and calyx lobes. Some species have two to four calyx lobes, while others have the five lobes more common to the family as a whole. Flowers are golden yellow to orange, or with orange to reddish markings.

Amsinckia tessellata (devil's lettuce, checker fiddleneck) is common throughout much of California, including the deserts. It frequents disturbed areas under 2,200 m, often in sand or clay soils. Devil's lettuce probably is a warning not to eat the nice green leaves of any *Amsinckia*, due to an abundance of alkaloids and nitrates toxic to both humans and livestock. The rough hairs of this species, as well as most others, are a skin irritant, as well as being a deterrent to harvest or ingestion. *Tessellata* means "checkered" in Latin, and refers to the arrangement of many bumps or warts on the backs of nutlets. *Amsinckia tessellata* is California's only desert species of Boraginaceae with yellow flowers and four sepals (sometimes fewer) plus coarsely hairy stems.

Amsinckia tessellata (devil's lettuce)

Cryptantha—This genus includes species with self-pollinating flowers that are as small as 1 mm across and never open. Because pollinators do not need to be attracted to the sex organs, self-pollinating flowers (*antha*) can afford to be "hidden" (*crypt-*). This reduces the amount of energy expended on temporary flower parts such as petals, and may help to keep fruits inconspicuous. Larger flowers in the genus, up to 12 mm across, tend to require pollination, and usually are perennial rather than annual. All but one species have a predominately white corolla; however, petal appendages around the opening of the tube or the tubular inside part of the corolla may be yellow. The odd species with an all-yellow corolla is *Cryptantha confertiflora*, a perennial shown on page 84.

There are 52 species of *Cryptantha* in California, 29 of which occur in our deserts. Most require a strong hand lens (about 20 power) or, preferably, a microscope to differentiate. Examination of several fruits is important because keys are often based on the numbers of nutlets that mature, which is variable, and on nutlet surface features, which depend on maturity. Petals and sepals also may require careful measurements and evaluation of details. Even if you do not want to identify

Cryptantha angustifolia (narrow-leaved forget-me-not)

Cryptantha to species, you can easily appreciate their differing growth forms, flower sizes, abundance, and variety of habitats.

Cryptantha pterocarya (wing-nut forget-me-not)

Cryptantha angustifolia (narrow-leaved forget-me-not) is "narrow" (*angusti-*) "leaved" (*folia*), but so are many other *Cryptantha* species. It is a common annual under 1,400 m, and has spreading bristles covering every green part. All four nutlets mature, with one becoming slightly larger than its siblings.

Cryptantha pterocarya (wing-nut forget-me-not) is a sparsely branched annual, with rough hairs but not many sharp bristles. Sepals become wide ovals, neatly encasing fruits as they develop. Nutlets are distinctive because three of them are fringed or have a "wing" (*ptero-*) around their margins, while the fourth lacks the fringe, and is just a plain old "nut" (*carya*).

Cryptantha tumulosa (New York Mountains cryptantha, Pinyon forget-me-not) is an uncommon perennial *Cryptantha* found only in California's northern and eastern desert mountains and nearby Nevada. It likes gravel or clay of granitic or limestone origin, and grows at elevations from 1,400 to 2,100 m. It has long dense hairs that are fairly soft, and the foliage is gray green. There are usually several stems, each with tight clusters of flowers at the tips obscuring the coiled formation. One or more of the nutlets fail to mature, and those that remain have a ridge down the middle of the back and lumps that look "hilly" (*tumulosa*).

Cryptantha utahensis (**scented forget-me-not**) has a lovely fragrance. It is a widespread annual in sandy to gravelly soils of both deserts under 2,000 m, and extends into Arizona and Utah (*utahensis*). The small (2–4 mm across) white corolla has yellow appendages. Only one nutlet, or sometimes two, matures.

Cryptantha tumulosa (New York Mountains cryptantha)

Heliotropium (**heliotrope**)—This is a large genus with 250 species, only two of which are native to California. The Greek name means "sun" (*helio-*) "turning" (-*tropium*), referring to the fact that many species flower near the time the sun is "turning" at the summer solstice. A distinguishing feature of the genus is the style attachment to the top of the ovary, which is common with many other families but unusual within Boraginaceae. Later, the style falls off and four (or two in nonnative species) nutlets form from deep lobes of the ovary. Most other features within the genus are variable, and the two desert species that we have in California represent different patterns.

Heliotropium curassavicum (**Chinese pusley**) grows throughout California, usually in somewhat moist saline soils and around irrigation ditches, where it spreads

Cryptantha utahensis (scented forget-me-not)

from a perennial root. One of the first collections of *Heliotropium curassavicum* was in Curaçao—hence the specific epithet. The name pusley is thought to have come from purslane or *Portulaca*, some of which have been used as herbs, possibly by the Chinese. *Heliotropium curassavicum* has coiled spikes of white flowers, typical of other genera in the family, such as *Cryptantha* and *Plagiobothrys*. However, stems and leaves of *Heliotropium curassavicum* are completely hairless and somewhat fleshy. The other desert species, *Heliotropium convolvulaceum*, is an annual lacking coiled spikes of flowers. Its flowers grow singly in leaf axils.

Pectocarya—This genus has 15 species; eight of them occur in California, all but one of which can be found in our deserts. The flowers and leaves of this annual look a lot like those of *Cryptantha*, but the fruits will quickly set you straight. The Greek name describes them as "comb" (*pecto-*) "nuts" (*carya*), referring to the nutlet fringes divided into teeth, though not nearly so regular as the teeth of a comb. Teeth are easily observable on the edges of nutlets that are spread widely in an X-formation.

Pectocarya setosa (**stiff-stemmed comb-bur**) can be as small as about 2 cm but is sometimes 10 times that. It has opposite leaves at the base of its stems, and alternate leaves above. The four spreading nutlets are oval to round but unequal, with three having wide membranous wings and the fourth having trimmed wings. Sepals become larger than nutlets and grow several stiff bristled hairs as well as numerous

Heliotropium curassavicum (Chinese pusley)

Pectocarya setosa (stiff-stemmed comb-bur)

shorter unarmed ones. *Setosa*, meaning "bristly," refers to sharp hairs and bristles on leaves, sepals, and stems.

Plagiobothrys (**popcornflower**)—This genus of 65 species is almost indistinguishable from *Cryptantha*. Only the nutlet attachment scar is reliable (almost) for making the distinction. In *Cryptantha* the scar is sunken, whereas in *Plagiobothrys* it is raised. Also, *Plagiobothrys* has more variability in the scar than *Cryptantha*, as reflected in the Greek name meaning "sideways" or "oblique" (*plagio-*) "pit" (*bothrys*). Species with a "sideways pit" have an angled scar, which sometimes is at the base of the nutlet and sometimes closer to the middle. Occasionally, it is on a short stalk. The backs of nutlets are usually heavily textured with ridges or tubercles, and often have a midrib. All of these nutlet features are important in species identifications. The common name conjures a less complex image, and we can think of the white petals, often with yellow appendages, as resembling buttery popcorn. There are 10 species of popcornflower in California's deserts, and 39 in the state as a whole.

Plagiobothrys arizonicus (Arizona popcornflower), young plant

Tiquilia plicata (plaited coldenia)

Plagiobothrys arizonicus (Arizona popcornflower) is a common annual with purple sap that will stain your hands. In older plants purple coloring can be seen in large veins and along the margins of leaves. There are usually just two mature nutlets per fruit, and they have a round scar near the middle of the inner nutlet edge. *Plagiobothrys arizonicus* grows in both deserts at elevations under 2,100 m, usually in shrubby areas with coarse soils.

Tiquilia—There are 27 species of *Tiquilia*, all in Western Hemisphere deserts, and four of them are in California's deserts. The genus name is taken from one of the South American names for a native species. Plants may be annuals, perennials, or subshrubs, and we have them all. Superficially, they look quite different from other Boraginaceae genera. Flowers are generally pink to purple or blue, and leaves are wide ovals rather than elongate and usually have sunken veins. Most species have opposite—or forked—branching, and spread across the ground. The reliable distinguishing feature is the deeply divided style, with two distinct branches, differentiating *Tiquilia* from all other genera of Boraginaceae. In California species fruits are four-lobed rather than divided into nutlets, also differentiating them from other common genera.

Tiquilia plicata (plaited coldenia) is a common species on dunes and sandy flats under 900 m. It is a perennial that feels slightly woody or at least stiff, even though the stems are thin and low. Leaves are whitish with tiny hairs, and have 4–7 pairs of deeply sunken veins. The pattern formed by venation appears "plaited" (*plicata*). Flowers grow in small clusters, and petals are lavender to bluish. Coldenia is a former genus name.

Examples and Uses

Most Boraginaceae members are unpalatable, inedible, or even poisonous—borage (*Borago officinalis*) being a notable exception. Borage, an annual, grows to a meter in height and has very hairy wide-oval leaves that are good in salads when very young or may be used as cooked greens. The bright blue flowers are an attractive garnish and also are edible. Borage is a common ingredient of herb gardens.

Several genera of Boraginaceae have widely known ornamental members, including Virginia bluebells (*Mertensia*), viper's bugloss (*Echium*), heliotrope (*Heliotropium*), and of course forget-me-not (*Myosotis*). The genus *Lithospermum* has been used by some North American Indian groups as a contraceptive, and *Symphytum* has been used as a comfrey, or wound healer. There is even a source of timber in the family (*Cordya*), and a red dye is made from roots of *Alkanna*.

BRASSICACEAE (CRUCIFERAE)

MUSTARD FAMILY

Mustards make wonderful spices, salad greens—and weeds. Many have successfully escaped from gardens to occupy a wide diversity of habitats, and a few are alarmingly well established in our deserts.

Brassica, meaning "cabbage" in Latin, is the root for the revised family name. The former name, Cruciferae, refers to the "cruci-form" or crosslike shape formed by the four petals of each flower.

Brassicaceae Icon Features

- Four free petals spreading above four sepals to form a cross or X
- Four long stamens plus two short stamens (sometimes two or four stamens in total)
- Superior ovary often long and thin with a short style

Distinctive Features of Brassicaceae

Look for four separate petals in the shape of a cross, with four long stamens and two short ones. There are rare exceptions: a few species have six uniformly sized stamens, or only two or only four, and a few species have no petals. Individual petals are usually in the shape of an L—narrow at the attached end and wider where they spread to form limbs of the cross. Typical colors are yellow, white, deep red, pink, or lavender. Flowers, though often small, are frequently displayed in large, elongated clusters on the ends of shoots and therefore appear quite showy.

The ovary is superior and divided down the middle by a membranous partition, known as the **septum**, that resembles waxed paper or cellophane.

Caulanthus inflatus (desert candle), Black Canyon area northwest of Barstow

This is readily visible as fresh fruit is dissected, and often becomes a distinctive feature of dried fruit that has split open. The (usually) many seeds may be seen attached to the perimeter rim of the septum as the sides of the drying fruit peel away starting at the stem end. Whole fruits are frequently long and thin, sticking out from the stem like bristles on a bottle brush. Several genera have short, rounded or triangular fruits, sometimes notched at the tip. Fruit shape, orientation of the septum with respect to any flattening of the fruit, and arrangement of seeds are important in determining the genus of a specimen.

Members of the Mustard Family are usually annuals, biennials, or perennials, with basal or alternate leaves, or both. A few species are shrubs or subshrubs. Leaves do not have stipules, but may be variously shaped from entire to deeply lobed or compound. Many species are hairy, and individual hairs may be branched in ways that help to identify species.

As the name "mustard" implies, most members of the family contain spicy oils, in a clear astringent sap.

Similar Families

The Caper Family (Capparaceae) is closely related to the Mustard Family, and may look similar for species that have four free sepals, four free petals, and six stamens. However, Capparaceae species never have four long stamens and two short ones, and they lack a septum partitioning the ovary chamber. In addition, Caper Family members generally have an ovary on the end of a stalk, which is rare in mustards. However, two genera of capers found in California's deserts have ovaries with a tiny to nonexistent stalk, and a few mustards have ovaries with a conspicuous stalk, so this characteristic is not utterly reliable. The most common desert genus of Capparaceae, *Cleome,* has flowers with bilateral symmetry, very unlike the characteristic cross of the Brassicaceae. Moreover, of the few species of Capparaceae represented in California deserts, all have leaves with three leaflets, a feature not found in any of our Brassicaceae species.

A few other families have members with four free petals, but they do not also have four long and two short stamens, superior ovaries, or fruits with a septum.

Family Size and Distribution

Worldwide, Brassicaceae is represented by 300 genera and 3,000 species, abundant in cool regions of the Northern Hemisphere. The greatest variety of genera occurs in southwest and central Asia and the Mediterranean region. The United States and Canada together have almost 100 genera and over 600 species.

California has 63 genera of Brassicaceae, almost half of which are entirely nonnative. Thirty genera inhabit California's deserts, nine of which consist only of nonnative species. This is a high proportion of weeds, and some of them are extremely widespread and well established. In general, however, the numerous weedy species of the mustard family are more successful where there is more moisture than in deserts.

Desert species occupy a wide range of habitats from flowing water to dry sand, and they cover virtually all elevations. At low elevations annuals, including weeds, bloom very early during cool weather and set seed rapidly, assuring future crops and possibly crowding out or stealing moisture from native annuals that grow later or more slowly. At high elevations perennials, which use seasonal dormancy to survive cold and drought, are more common.

California Desert Genera and Species

The more conspicuous flowers of Brassicaceae are typically yellow, noticeable mainly due to their abundance, while smaller or less dense flower clusters may be dark reddish, purple, pink, or white. Native California desert representatives of the family are generally annuals or biennials, often with a dense cluster of basal leaves and a few smaller leaves on tall stems bearing numerous yellow or white flowers. Desert annuals of the family, especially nonnatives, tend to sprout readily along roadways or in other disturbed locations with a bit of extra runoff water. Dried skeletons of stems and flowering stalks are a common sight late in the season following a wetter than usual spring.

Growth forms in the desert range from annuals that spread prostrate across the ground to shrubby-looking perennials. Subshrubs and shrubs are rare, upright annuals common.

SIMPLIFIED KEYS TO DESERT GENERA OF BRASSICACEAE

We separate desert Brassicaceae genera into three groups for the purpose of simplifying keys. The distinction can be made easily on the basis of fruit shape. Find the correct group in the key below, then follow the appropriate group key. The only genus with an ambiguous fruit, *Draba*, can be found by using either of two keys.

Group 1
Fruit long
Modest tansy mustard (*Descurainia sophia*)

Group 2
Fruit short and thin
Woolly-fruited peppergrass
(*Lepidium lasiocarpum*)

Group 3
Fruit short and fat
Palmer bead pod (*Lesquerella tenella*)

The shape of Mustard Family (Brassicaceae) fruits is used to separate genus keys into three different groups.

KEY TO GROUPS

1. Fruit long (clearly longer than wide by at least 3 times), generally round in cross section, but sometimes flattened .... GROUP 1
1' Fruit short (about as wide as long).
  2. Fruit thin (flat like a pancake, some with a small bulge) .... GROUP 2
  2' Fruit fat (spherical or inflated looking) .... GROUP 3

KEYS TO GENERA BY GROUP

Both flowers and mature fruits are needed to identify Brassicaceae genera, and flowering stems usually have opening flowers near the top and maturing fruits near the bottom, providing a continuous supply of material suitable for examination. Both sepal and petal color are useful, but it is the shape of fruits and orientation of the septum that are most true to genus. When fruits are flattened, the septum may be parallel to the broad walls of the fruit, having essentially the same outline as the fruit, or it may be perpendicular to the walls of the fruit. In the latter case, the septum is narrow, attaching the two flat fruit walls down the middle. The location of attachment usually shows as a faint line distinguishing two halves of the flat fruit surface. Some long fruits have a pointed extension, or **beak**, at the tip, resembling a bird's beak. Although leaf shape tends to be highly variable, even on a single plant, information about the larger basal leaves is sometimes a good indicator of genus.

GROUP 1 KEY: Fruit long.

1. Sepals not green, often yellowish or purplish.
   2. Fruits round in cross section.
      - Fruit with a long threadlike stalked base. Sepals spreading widely or reflexed (2 spp) .... *Stanleya*
      - Fruit cylindrical throughout, not stalked. Sepals usually erect, forming a vaselike enclosure for the lower portions of petals (9 spp) .... *Caulanthus*
   2' Fruits somewhat flattened in cross section.
      3. Fruit flattened perpendicular to septum, sides of fruit bulging. Sepals spreading open at flowering. Petals yellow, some tinged purple .... ***Tropidocarpum gracile***
      3' Fruit flattened parallel to septum. Sepals remaining in vase formation. Petals variously colored.
         4. Fruit with stalked base 1–3 mm long. Petals white, 6–9 mm .... *Thelypodium integrifolium*
         4' Fruit without stalked base. Petals purple, or whitish to yellowish often marked with purple.
            - Petals purple. .... ***Streptanthus cordatus***
            - Petals whitish or yellowish with purple veins .... ***Streptanthella longirostris***
1' Sepals green.
   5. Plants growing out of water or soft mud of springs or streams (2 spp) .... *Rorippa*
   5' Plants occupying dryer habitats.
      6. Annual with hollow stems. Petals white to pale yellow, 3–6 mm .... *Guillenia lasiophylla*
      6' Not an annual with hollow stems, but often annual. Petals white to yellow, some purplish or pinkish.
         7. Petals generally white, some pinkish to purplish.
            8. Fruit flat in cross section.
               - Fruit narrow, about 10 times longer than wide. Leaves entire or toothed (6 spp) ... ***Arabis***
               - Fruit narrow, about 10 times longer than wide. Leaves deeply lobed (2 spp) .... ***Sibara***
               - Fruit wide, about 3 times longer than wide. Leaves entire to toothed (2 spp) .... *Draba*
            8' Fruit round in cross section.
               - Fruit very hairy, not beaked. Petals white, 4–5.5 mm .... *Halimolobos jaegeri*

• Fruit not hairy, tip with flat beak. Petals white to slightly yellowish, 15–20 mm .... *Eruca vesicaria*

• Fruit not hairy, tip with a narrow round beak. Petals white, 3–6 mm .... *Guillenia lasiophylla*

7' Petals yellow, some pale yellow or orange-yellow.

  9. Leaves fernlike, all subdivided into small toothed or lobed leaflets (4 spp) ...... **Descurainia**

  9' Leaves entire, toothed, or deeply lobed, but few if any subdivided into separate leaflets.

    10. Basal leaves deeply lobed.

      11. Fruit 10–15 mm, held upright parallel to stem .... *Hirschfeldia incana*

      11' Fruit 30 mm or more, spreading or rising but not parallel to stem.

        • Basal leaves with broad pointed lobes, upper leaves with linear lobes (3 spp) .... **Sisymbrium**

        • Basal leaves with deep rounded lobes, wide at tip, few small upper leaves .... **Brassica tournefortii**

    10' All leaves entire to barely toothed.

      • Leaves narrow. Petals deep yellow to orange-yellow .... **Erysimum capitatum**

      • Leaves broad. Petals lemon yellow to pale yellow .... *Conringia orientalis*

GROUP 2 KEY: Fruit short and thin.

1. Septum parallel to broad fruit walls, or fruit with a single chamber (septum apparently absent).

  • Septum parallel to broad fruit walls. Fruits linear or oval with pointed ends, about 3 times longer than wide. Petals white or absent (2 spp) .... *Draba*

  • Septum not apparent, fruit with a single chamber and single seed. Fruit round with a broad flat, membranous rim (2 spp) .... **Thysanocarpus**

1' Septum perpendicular to flat fruit walls.

  2. Fruits with 2 distinct halves, not round or oval in overall shape.

    • Fruit with 2 round halves, like eye glasses. Petals white to lavender .... **Dithyrea californica**

    • Fruit heart-shaped, like an elongated valentine. Petals white .... *Capsella bursa-pastoris*

    • Fruit guitar-shaped or lyre-shaped, wavy. Petals brownish to purplish .... **Lyrocarpa coulteri**

  2' Fruits round or oval in overall shape, often notched at the tip, some with partially winged rims.

    • Fruit oval, not notched at tip. Petals white, 1 mm .... *Hutchinsia procumbens*

    • Fruit broadly oval, notched, winged all around except for notch at tip. Petals white, 3–4 mm . ... *Thlaspi arvense*

    • Fruit round to oval or slightly angular, notched, occasionally winged near tip. Petals, if not absent, white, less than 3 mm; occasionally yellow (10 spp) .... **Lepidium**

GROUP 3 KEY: Fruit short and fat.

1. Petals yellow.

  • Leaves silvery hairy, clumped at base. Fruit hairy, lumpy halves forming a misshapen sphere .... *Physaria chambersii*

- Leaves green hairy, several at base, many along stem. Fruit hairs sparse to none, smooth halves forming a sphere or egg shape (2 spp) .... *Lesquerella*

1' Petals white.
- Leaves toothed. Fruit a slightly flattened sphere, surface smooth .... *Cardaria pubescens*
- Leaves deeply lobed. Fruit with 2 partly spherical halves (like a balloon with a string constricting the middle), surface with short spines (2 spp) .... *Coronopus*

Arabis (rock-cress)—*Arabis* means "from Arabia," but all of the 39 species found in California are native. There are six desert species, the most widespread of which is *Arabis pulchra*. Desert species of *Arabis* are perennials of small stature (less than 80 cm), sometimes with a woody caudex and most leaves clustered near a sparsely branched base. Our desert species have branched hairs on leaves and often on stems and fruits. Flowers are small (less than 1 cm) and either white or pink to deep reddish or purplish. Fruits are long and thin with many seeds, and flattened parallel to the septum.

Arabis perennans (perennial rock-cress)

Arabis pulchra (prince's rock-cress)

Arabis perennans (perennial rock-cress) is found in canyons and on gravelly slopes of desert mountains. It is a small (20–60 cm) "perennial" (*perennans*) with hairy basal leaves and a few delicate flowering stems that are generally hairy below but hairless above. There are several pinkish lavender to purple flowers, about 5 mm across, near the tip of each stem or branch, with long (4–6 cm), thin fruits forming on 1–2 cm pedicels below the older flowers.

Arabis pulchra (prince's rock-cress), with a specific epithet meaning "pretty," is found on slopes and in canyons and washes throughout California's deserts, as well as in Utah, Arizona, and Mexico. It has a woody caudex and may be branched a little above ground level. Narrow leaves, clustered near the base, are much diminished along the stem. Flower petals are deep reddish purple and shaped like a bent spoon. Sepals grow erect and close together, enclosing the thin basal ends of the four petals, while the wider bowls of the spoon-shaped petals spread to form a cross. There are five varieties of *Arabis pulchra*, three of which are found in California deserts.

Brassica (mustard)—All five *Brassica* species found in California are annuals from Europe or Asia. Only one species, *Brassica tournefortii*, is naturalized throughout California's deserts, but others are occasionally found as weeds near habitations.

Brassica tournefortii (**Asian mustard**) is abundant in disturbed areas, although it is not restricted to them. It is becoming extremely common in California's deserts and may be adversely affecting native vegetation. It grows vigorously in the early season in response to very little water and sets seed rapidly. Under favorable conditions it grows to about 60 cm, with long (25 cm) basal leaves that are lobed to the mid-vein and toothed. Lower stems and under surfaces of leaves develop spiny hairs that discourage weeding. Branching upper stems have very reduced to absent leaves and are tipped with enough yellow flowers to make numerous long, cylindrical fruits. The specific epithet honors Joseph Pitton de Tournefort (1656-1708), a French botanist who is credited with being first to properly define genera.

Brassica tournefortii. (Asian mustard, African mustard, Tournefort's turnip)

Caulanthus (**jewel flower**)—*Caulis* refers to the "stem" of a plant and *anthus* means "flower," so *Caulanthus* means "stem-flower." This description fits the most distinctive member of the genus, *Caulanthus inflatus*, which has an enormously fat stem with little flowers and long fruits growing directly out of it. California has nine desert species of *Caulanthus*, both perennials with a woody caudex and annuals, occurring mainly in the Mojave Desert. Most leaves are basal; the few leaves along the stem are much smaller and their bases clasp partway around the stem. *Caulanthus* flowers are slightly asymmetric, with petals somewhat paired instead of strictly in a cross shape. It is common for petals to be dark purplish or brownish, or to have dark-colored veins, and usually the edges of the petals are dry and translucent as if aging. The four sepals, which tend to be dark-colored too, or at least not green, grow with their edges close together forming a vaselike enclosure over the narrow bases of petals. Sometimes the base of each sepal has a small pouchlike bulge.

Caulanthus crassicaulis var. *crassicaulis* (thick-stemmed wild cabbage)

Caulanthus crassicaulis (**thick-stemmed wild cabbage**) has an impressively "thickened" (*crasse*) stem. Although similar in this respect to *Caulanthus inflatus*, *Caulanthus crassicaulis* is easily

Caulanthus inflatus (desert candle)

Descurainia pinnata (yellow tansy mustard)

distinguished by being a perennial with a woody caudex, by its deeply lobed basal leaves, and by the grayish green color of the stem. Two varieties of *Caulanthus crassicaulis* are found in north and central desert mountains.

Caulanthus inflatus (**desert candle**), with its straight, "inflated" (*inflatus*) stem standing to as tall as 1 m, looks like a bright yellow-green candle. Easy to spot in open sandy stretches of the southwestern Mojave Desert, it is an annual with wide oval leaves on lower portions of the fat stem.

Descurainia (**tansy mustard**)—This genus is named after the French apothecary François Descourain (1658–1740), whose profession, at the time, was closely tied to botany. Four species are found in California's deserts, one of which is nonnative. Plants in this group are fairly delicate looking, with many very small flowers (about 3 mm), lots of thin cylindrical fruits on thin pedicels, and intricately divided leaves sometimes looking almost fernlike. Petals are yellow to whitish and not much bigger than sepals. The length of fruits and angles at which they spread from the stem are important in identifying species.

Descurainia pinnata (**yellow tansy mustard**) is a hairy annual with finely divided leaves that are "feathered" (*pinnata*). Each of the ovary chambers contains two rows of seeds, and fruits are usually club-shaped. Four subspecies inhabit California, three of which occur in the desert, frequently in dry streambeds or washes.

Dithyrea (**spectacle-pod**)—You will recognize this genus immediately from the shape of the fruit (pod), which has two almost-round halves joined together like eyeglasses (hence the common name). Some botanist thought they looked like two shields, which is what *di-thyrea* means in Greek. A distinguishable rim defines the edges of the glasses, and each half holds just one seed. There are only two species in California, one (rare) at the beach, and one in the desert in sandy locations at low elevation.

Dithyrea californica (**California spectacle-pod**) is an annual with fragrant light cream to light lavender flowers. Each of the four petals is a little more than 1 cm long and has three prominent veins. Plants grow to about 30 cm and may be abundant in sand dune areas after winter rains.

Erysimum (**wallflower**)—The Greek root for *Erysimum* refers to medicinal uses of mustards to draw away blisters. There are over 160 species in the Northern Hemisphere, but only a few in California, where each of the natives is evolved from the widespread *Erysimum capitatum*. The subspecies *capitatum* is found in the desert as well as almost everywhere else in California.

***Erysimum capitatum* ssp. *capitatum* (western wallflower)** is a showy yellow to somewhat orange flowered biennial or perennial. Flowers form an almost spherical cluster at the end of the stem—hence the Latin name *capitatum*, meaning a "knob-shaped head." The petals are in the shape of an L, with the long end up to 3 cm and the rounded short end about 1 cm or less. Fruits, which spread upward below the flowers, are long (3–15 cm) and cylindrical. They feel somewhat four-sided but may be slightly flattened parallel to the septum. There is usually only one stem, with narrow leaves that may be slightly toothed. Leaves extend up the stem to fruits but are bigger and more dense near the bottom.

Dithyrea californica (California spectacle-pod)

***Lepidium* (peppergrass, pepperwort)**—*Lepidium* is from the Greek meaning "small scale," which Edmund Jaeger in *Desert Wild Flowers* (1940) speculated might refer to the use of some species in treating diseases such as leprosy that form scales on skin. Other interpretations suggest the name refers to small scale-like fruits. The genus has around 175 species worldwide, including 10 in California's deserts, ranging in form from low spreading annuals to shrubby perennials. Petals of *Lepidium* are small and usually white, but sometimes are absent, and there may be four or two stamens instead of the usual six that are characteristic of Brassicaceae members. Fruits are rounded, heart-shaped, or elliptical and are flattened perpendicular to the septum, with only one seed per chamber.

***Lepidium flavum* (yellow peppergrass)** is—as the Latin -*flavum*, meaning "yellow," implies—an exception to the white-flower "rule" of the genus. This annual branches from the base to spread across the ground,

Erysimum capitatum ssp. *capitatum* (western wallflower)

Lepidium flavum (yellow peppergrass)

Lepidium fremontii (desert alyssum)

often in large patches: a single plant can be a half meter or so across. Yellow flower petals are only 2–3 mm long, but, in abundance, they are colorful enough to be seen from a distance. This species has six stamens. The variety *felipense* (Borrego Valley peppergrass), considered rare in California, is confined to sandy soils of the Borrego Valley. The variety *flavum* is widespread in California deserts on alkaline soils and flats.

Lepidium fremontii (**desert alyssum**) is named after the famous explorer of America's Southwest, John C. Frémont. It is a shrubby perennial with white flowers that are almost always attended by many small flies. The plant is generally less than 1 m tall and is densely branched, with small (3 mm) flowers clustered near the ends of numerous branches. The flowers and flattened, rounded fruits resemble sweet alyssum (*Lobularia maritima*), also in Brassicaceae, accounting for the common name. Leaves are linear and deeply divided into several very narrow lobes. They are inconspicuous when the plant is blooming, which may be following any good rain. There are two varieties in California's deserts.

Lesquerella (**bladderpod, bead-pod**)—This genus is named after Leo Lesquereux (1806–1889), an authority on plant fossils. There are roughly 95 species, concentrated in North America, only two of which occupy California's deserts. *Lesquerella* species are usually biennials or perennials, and they are densely hairy, often taking on a silvery appearance. They have six stamens and yellow flowers. Seed pods tend to be round and fattened into a sphere, with fewer than a dozen seeds per chamber.

Lesquerella tenella (**Palmer bead-pod**) is an annual growing to about 60 cm, with branched stems and clusters of yellow flowers at stem tips. Fruits are almost spherical, like beads on the ends of S-curved pedicels. A persistent style 2–4 mm long remains on fruit tips. Leaves 3–6 cm long and narrowly oval grow at the base of plants, with smaller ones continuing along stems. *Tenella* means "delicate" or "tender." *Lesquerella tenella* likes sandy soils at

Lesquerella tenella (Palmer bead-pod)

elevations under 1,200 m.

Lyrocarpa (lyre-fruit)—The Greek name refers to the unusual "lyre-" (*lyro-*) or guitar-shaped "fruits" (*carpa*) found in this small genus. There are only three species, all found in southwestern arid regions of North America, with one growing in California.

Lyrocarpa coulteri var. *palmeri* (Coulter lyre-fruit)

Lyrocarpa coulteri var. *palmeri* (Coulter lyre-fruit) is named after the Irish botanist Thomas Coulter (1793–1843), who is said to have "discovered," botanically speaking, the Colorado Desert segment of the Sonoran Desert in 1831. The varietal name is that of another botanical explorer of the southwestern United States and Mexico, Dr. Edward Palmer (1831–1911). *Lyrocarpa coulteri* var. *palmeri* is a perennial with a somewhat woody caudex and irregular branches, normally about 50 cm tall. Flowers of this species are just as unusual as the fruits. Petals are long (up to 2.5 cm) and narrow (1–3 mm wide), usually twisting a bit and tapering to a narrow point; their color is rust brown to deep purple. This variety, considered uncommon in Cali-fornia, is found in canyons along the southwestern border of the Sonoran Desert.

Rorippa (water cress)—This genus likes water: all 11 species occurring in California require moist habitats, although not necessarily open water such as that utilized by our salad favorite, *Rorippa nasturtium-aquaticum*. There are about 75 species in all, widely distributed across the globe, and the genus name is Old Saxon, probably addressing a broader group of edible mustards. Flowers are usually small, with petals either absent, yellow, or white. Leaves—the tasty part—may be either entire or divided into separate lobes. California's deserts have one species with yellow petals and one with white.

Rorippa nasturtium-aquaticum

Rorippa nasturtium-aquaticum (water cress)

Sisymbrium irio (**London rocket**)

(**water cress**) is frequently abun-dant in freshwater streams and springs around the world. It is also cultivated, found in grocery stores, and admired for its peppery flavor in salads and sandwiches. *Nasturtium* means "twisted nose" in Latin, referring to the flavor; *aquaticum* means "of the water." (Nasturtiums of the land—or at least of gardeners—are in the family Tropaeolaceae.) Leaves are divided into three to seven almost round leaflets. Stems are partly submerged, partly emergent, and usually 30–50 cm long, with most leaves above the water line. Desert populations are found mainly in springs, where they may have been introduced as a food crop, although they are native to the state.

Sisymbrium—Many of the 90 members of this genus, with a Greek name meaning "various mustards," are widespread weeds, occurring throughout the world. All three of the species present in California's deserts came originally from Europe. They are annuals or biennials that like disturbed areas and grow rapidly from seed. Leaves of several species are used as salad greens.

Sisymbrium irio (**London rocket**) is one of the salad members of the genus, with long lobed basal leaves and smaller ones closer to flowers and fruits. Flowers are only 3–4 mm long and light yellow, arranged in dense clusters at the tips of branches, with long thin fruits maturing from the older flowers below. London rocket likes roadsides and other disturbed areas under 800 m. Its specific epithet comes from the Greek word for rainbow.

Stanleya (**prince's plume**)—Edward Stanley (1773–1849), an English ornithologist, was at one time president of the Linnean Society of London, where the original plant collections of Carl Linnaeus are housed. The common name, prince's plume, is descriptive of the long, tall, full-flowering shoots of this genus. The plume effect is enhanced by extra-long stamens that extend well beyond the conspicuous petals. The approximately equal length of the six stamens and the fact that the fruit forms on a stalk above the receptacle led early botanists to classify these plants in the Caper Family (Capparaceae). However, the septum of the fruit—a more reliable indicator than stamen lengths and fruit stalks—is consistent with Brassicaceae. *Stanleya* has only six species, all of which are found in the western United States, frequently on soils rich in selenium. Although trace amounts of selenium are important in the human diet, *Stanleya* may concentrate selenium to toxic levels and should not be eaten.

Stanleya pinnata (**prince's plume**) has several to many branches from a woody base and often grows to 1.5 m or more. In full bloom it is a striking plant, visible from a considerable distance in open sandy areas or desert shrubland under 1,850 m. Flowers of *Stanleya pinnata* form dense spikes up to 50 cm long. Petals are 1–2 cm long, and stamens even longer. Long wavy hairs on the interior lower ends of each petal are a good distinguishing feature of the species. Also, lower leaves are deeply pinnately lobed. *Pinna* means "feather" in Latin, and *pinnata* refers to the veins and lobes of the leaf,

Stanleya pinnata (prince's plume)

which are arranged like the ribs of a feather. The other species in our deserts, *Stanleya elata*, has large oval leaves that may be toothed but are not lobed. It is found mainly in the canyons of our desert mountains at 1,300–2,000 m.

Streptanthella (**twist-flower**)—This is a genus with only one species. *Streptanthella* is the diminutive form of *Streptanthus*, the name of a much larger genus of Brassicaceae (but one with minimal desert representation). *Strept-*, meaning "twisted" in Greek, combined with *anthus*, meaning "flower," plus -*ella*, the diminutive ending, adds up to "small twisted flower." Flower petals are 3–4 mm with wavy margins, but "twisted" really overstates the case.

Streptanthella longirostris (**long-beaked twist-flower**) is common in sandy soils throughout California's deserts. It is an annual, usually with multiple branches from a single stem and narrow wavy margined leaves, largest (3–6 cm long) near the base. It may have either tiny simple hairs on its lower portions or no hairs at all. The color of leaves and stems tends toward gray-green due to a fine powdery white coating. Several small flowers grow along upper portions of the main stems, spaced about 1 cm apart. Sepals, only 2–3 mm long, form a greenish purple to purple cylinder that almost hides the narrow white to yellowish petals. The four longer of the six stamens usually extend a bit past the petals. The long (35–45 mm) and thin (1.5 mm) fruits hang in pendant postures and are somewhat flattened parallel to the septum. The name *longirostris*, or "long beak," refers to a narrow beaklike tip on each of the fruits.

Streptanthus (**jewelflower**)—There are roughly 40 species of *Streptanthus*, all in southwestern North America, but only one of them

Streptanthella longirostris (long-beaked twist-flower)

Streptanthus cordatus (heart-leaved twist-flower)

Thysanocarpus laciniatus (lacepod)

occurs in California's deserts. Most members of the genus have non-green sepals shaped like a vase embracing wavy-edged petals: hence the name, meaning "twisted" (*strept-*) "flower" (*anthus*). Fruits are long and flattened parallel to the septum.

Streptanthus cordatus (**heart-leaved twist-flower**) is restricted to higher elevations (1,400–2,800 m) of our eastern desert mountains. Its flower, which is almost bilaterally symmetrical, is dominated by a maroon calyx. Narrow petals also are maroon. It is named "heartlike" (*cordatus*) for the shape of its leaves, whose wide lobed base—resembling the top of a valentine—wraps directly around the stem, without benefit of a petiole. Plants are perennial and sometimes grow to almost 1 m tall, with few or no branches.

Thysanocarpus (**lacepod, fringepod**)—Four of the five species in this genus live in California, all natives. Two species occupy desert sites, succeeding because of their annual habits and their ability to respond rapidly to early rainfall. Fruits are almost round, very flat, and with a membranous rim that sometimes is perforated and lacy. In Greek, *thysan* means "fringe" and *carpus* is a "pod" or "fruit." Breaking family rules, members of this genus have no septum, and the fruit has just one chamber and one seed.

Thysanocarpus laciniatus (**lacepod, fringepod**) is common in rocky areas under 2,400 m, and does best with a bit of shade. Generally, there are few leaves at the base, with more along stems. Leaves on a single plant range from deeply divided into narrow lobes to linear—hence the specific epithet, meaning "slashed into narrow segments."

Tropidocarpum—This small genus has only two species, one of which is probably extinct. *Carpum* means "fruit," and "*tropido-* describes it as "keeled," or having ridges along the sides where it will open when dry.

Tropidocarpum gracile is found in the western Mojave Desert in grassy, open habitats such as stream banks or pastures. It is a "graceful" or "slender" (*gracile*) annual, growing to a height of up to 50 cm, usually with some branching near the base and leaves that are deeply lobed. Leaves, though larger and more abundant near the base of the plant, grow sparsely all the way up the stem. Bracts where fruits and flowers join the stem look like miniature leaves. There may be a few simple hairs on stems, leaves, or fruits. Petals are only about 4 mm long and are yellow, sometimes with a purple tinge.

Examples and Uses

Most of us remember the vegetable that President George Bush I refused to eat! In fact, broccoli has numerous relatives that, in spite of differing personal preferences, are normally considered edible. Cultivation of the ancestral colewort (*Brassica oleracea*) has amplified various parts of the plant into foods we find in our grocery stores: broccoli and cauliflower are enlarged flower clusters (eaten in the bud stage); brussels sprouts are enlarged buds grown along a tall stem; kohlrabi is formed from a swollen stem and leaf bases; and cabbage is densely packed leaves growing on a condensed stem. All are usually considered the same species as their wild ancestor, even though they appear very different.

The strong oils typical of the Mustard Family are largely responsible for strong flavors in many other foods with mixed reputations, such as horseradish (*Armoracia*), turnips (*Brassica*), radish (*Raphanus*), and water cress (*Rorippa*). As Mustard Family foods are cooked, the oils and other chemicals break down and recombine to form unpleasant

Tropidocarpum gracile

sulfur compounds, which not only tend to smell bad but also make the flavor stronger as cooking time increases. Minimal cooking often gets the best results from mustards.

Canola oil is extracted from seeds of *Brassica* species. Canola is a refined and edible version of rapeseed oil, which is used for industrial purposes.

Mustard that we use as a condiment is made from the seeds of white mustard (*Sinapsis alba*), Indian mustard (*Brassica juncea*), and black mustard (*Brassica nigra*), which are cracked and wetted to release the flavors. Although these mustards are native to Eurasia, they grow widely throughout California. According to some stories, black mustard seeds were scattered by the Spanish padres so that the tall yellow-flowering shoots would mark the route between missions along the California coast.

Many Brassicaceae species are cultivated as ornamentals. The most familiar include sweet alyssum (*Lobularia maritima*), candy-tuft (*Iberis*), and wallflower (*Erysimum*).

Opuntia erinacea (Mojave prickly-pear), Panamint Mountains

CACTACEAE

CACTUS FAMILY

Cactus is symbolic of the American desert. Dangerous spines, hugely fattened stems, and dramatically colored flowers are common images of the Cactus Family. Many species are so interesting and beautiful that they are collected illegally from the wild, almost to extinction. The marvelous adaptations of cacti to seasonal drought include storage of water in the expanded stem, necessitating its protection with spines to resist thirsty animals. The stem is also important for photosynthesis because there are no leaves, and cacti are able to employ alternative forms of photosynthesis depending on environmental conditions.

The Greek *cact* means "prickly" or "thorny plant" and has been adopted as the basis for both the scientific and common family names.

Cactaceae Icon Features

- Green stems fattened, fleshy, and leafless
- Spines attached in areoles often with tiny sharp glochids
- Flowers with numerous petals
- Inferior ovary embedded in stem tissue

Distinctive Family Features

The Cactus Family is characterized by enlarged leafless stems with **areoles**—small pads or pockets occurring in a regular pattern over the stem surface, usually bearing one or more spines. Areoles are a specialized form of the bud that grows in leaf axils. However, in Cactaceae members, leaves (if any at all) are very short lived, occurring only on tiny new growth while there is still water available to the roots. As the stem expands, leaves fall off, and axillary buds develop into spiny areoles. Only Cactus Family members have areoles.

Cactaceae flowers are notable for their numerous stamens and petals, attached in a spiral arrangement. The outermost petals are usually greenish and transition into being sepals. To avoid having to make a distinction, petals and sepals collectively are called the perianth. The stamens and perianth base are fused together, forming a **hypanthium** that sits on top of an inferior ovary. Usually, the hypanthium is cuplike and a source of nectar for attracting pollinators.

The ovary is deeply submerged into stem tissue that looks like part of the fruit but generally has several to many areoles, often with spines. In *Opuntia* species the areoles have lots of nasty, tiny barbed spines called **glochids**. These are what stay in your fingers after you have attempted to pick some smooth-looking fruits.

There is just one style attached to the ovary, but the stigma may have several lobes and is often brightly colored, contrasting with the perianth color, making a fine bulls-eye for pollinators. In some species, the motion of an insect pollinator within the flower causes stamens to bend toward the center so that pollen-bearing anthers are sure to dust the insect as it exits. When pollen is being dispersed, the stigma is immature and unreceptive; pollen dusted onto the insect is thus carried away to fertilize a different flower that has a mature stigma.

Spines occur on most cacti and can be very dense. In addition to keeping herbivores at bay, they are thought to collect sparse rainfall and dew and direct it to the roots. Spines also help screen the sun's rays and keep stem temperatures within tolerable limits.

The shape of cactus stems and the numbers and sizes of spines are important features in identifying genera and species. The unusual shapes of cactus stems and their role in storing water require a special type of woody supporting structure. This can be seen in rotted cactus stems when the fleshy tissue has decomposed, exposing a network of wood looking something like a tube made from woody chicken wire. Judging from the weight of the branching networks supported by many species of cacti, the wood structure, although appearing light, is extremely strong.

Flower color is often useful in identifying cacti, but in some closely related species it is variable and may be a source of confusion. Some of the special vibrancy of cactus flowers is due to unusual types of nitrogen-containing pigments called betalains, present in Cactaceae members and several closely related families such as the Purslane Family (Portulacaceae). Although these pigments cannot readily be identified in the field, they have been useful in the laboratory in determining relationships among families.

Similar Families

The fattened leafless stems and spines of Cactaceae are mimicked by some members of the Spurge Family (Euphorbiaceae), but none of these look-alikes is native to California. You are likely to see them in landscaped desert areas, where they can be distinguished from cactus by the absence of areoles at the base of spines.

A few families possess desert adaptations similar to those of Cactaceae, including the Stonecrop Family (Crassulaceae), the Fig-Marigold Family (Aizoaceae), and the Purslane Family (Portulacaceae), but here the emphasis is on water storage in enlarged succulent leaves rather than stems, and none of the very few desert species in these families has spines or otherwise resemble any cactus of California's deserts.

Occasionally, ocotillo (*Fouquieria splendens*), in Fouquieriaceae, is mistaken for a cactus because of its numerous spines along branches that are often leafless. However, the branches of ocotillo are woody sticks that do not expand to store water, and there are no areoles at the bases of spines. When ocotillo is blooming, it is clear that the flowers are entirely different from those of cacti: the red petals of ocotillo are fused into a slender tube, as opposed to the numerous separate petals of Cactaceae. When in bloom, too, ocotillo stems usually are covered with small green leaves and no longer resemble cacti.

Family Size and Distribution

Cactaceae includes 93 genera and roughly 2,000 species, mainly in tropical and subtropical portions of the Western Hemisphere. Some species range well into temperate zones and up to elevations of 3,000 m. Although we think of cacti as desert species, some thrive in rain forest habitats, several are coastal, and still others have enormous ranges from deserts into high mountains.

Opuntia is the largest genus of Cactaceae, with about 200 species, some of which have spread from their American origins to other arid parts of the world. In Australia, the introduction and spread of *Opuntia* resulted in loss of native vegetation and widespread destruction of grazing lands.

California has nine genera of Cactaceae; eight of which are represented in our deserts, with the ninth rare on the coast near San Diego and on the Channel Islands. Almost all California Cactaceae species are natives and grow only in southwestern North American deserts and nearby arid zones. Some species, like giant saguaro (*Carnegiea gigantea*), have a very limited range in California, occurring only in the southeasternmost portion of the state. Saguaro are more abundant in the Sonoran Desert east of California, where there is more summer rainfall. Other cactus species, such as beavertail cactus (*Opuntia basilaris*), grow throughout desert slopes as well as adjacent areas, including the southern and eastern Sierra Nevada.

In general, cacti require fairly reliable seasonal rains that can be absorbed by shallow roots. Extremely dry places, like Death Valley, do not have many cacti. Good drainage is also required, frequently limiting cacti to rocky slopes or loose gravelly soil rather than fine clay flats.

California Desert Genera and Species

All California cacti are fleshy perennials that are leafless except on the youngest new growth. Almost all of our cacti have readily visible spines, and the few that do not are armed with microscopic barbs equally capable of inflicting pain and suffering. Although the flowers are the most attractive aspect of Cactaceae, it is mainly characteristics of stems and spines that are used to identify species. In fact, many cacti bloom in the summer when it is too hot for most people to be out looking for flowers.

SIMPLIFIED KEY TO DESERT GENERA OF CACTACEAE

Cactus stems are described by several key terms. **Jointed** means having segments that look as if they were made separately and then joined together.

Ribs are longitudinal furrows and ridges in the stem, so that if you cut it crosswise, the shape would be a multipointed star. **Tubercles** are small bumps or wartlike bulges of the stem that have an areole at the tip where the spines originate. Tubercles can be lined up along ribs, or can form spiral patterns on stems without ribs.

1. Stems jointed. Glochids with or without spines at areoles (13 spp and recognized hybrids) .... *Opuntia*

1' Stems not jointed. Glochids absent but areoles always with spines.

  2. Stems not or barely ribbed, tubercles prominent. Plants low growing (under 30 cm).

    • Wide-spreading spines radiating around 1 central spine, not hooked .... *Escobaria vivipara*

    • Small spreading spines plus 1–4 large central spines with hooked ends (3 spp) . *Mammillaria*

  2' Stems prominently ribbed, with or without tubercles. Plants various heights.

    3. Plants tall (often over 1 m). Stems generally single with zero to a few prominent branches.

      • Over 3 m, often branched, creamy white flowers open at night .... *Carnegiea gigantea*

      • Usually about 1 m, unbranched, yellow flowers open in the day ...... *Ferocactus cylindraceus*

    3' Plants low (less than 60 cm) with either single or clustered stems with zero or obscure branching.

      4. Stems generally single, about 15 cm tall, and ovoid to cylindrical, with prominent ribs and tubercles. Flowers greenish yellow, pink, rose-purple, to magenta (2 spp) .... *Sclerocactus*

4' Stems mostly clustered or mound forming.
- Stems bowling ball size or taller (to 60 cm), in clusters. Stout spines often reddish; indistinct tubercles. Tops of stems with cotton balls from old fruits. Flowers yellow .... *Echinocactus polycephalus*
- Stems cylindrical or spherical, about 10 cm in diameter, growing in mounded clusters. Spines straight or curved but not hooked. Flowers magenta to lavender or bright red (2 spp) .... *Echinocereus*

Echinocactus polycephalus var. *polycephalus* (cotton-top cactus)

Echinocactus (cluster barrel cactus, cotton-top cactus)—*Echinos* is a Greek word for hedgehog and sea urchin, animals with spines. So "hedgehog" plus "cactus" describes a cactus with spines—not a very distinctive feature within the family. There are just six species of *Echinocactus*, restricted to the southwest United States and Mexico, and only one species grows in California. All members of the genus have very woolly fruits that leave dried bundles of hairs, looking like balls of cotton, stuck among the dense spines on top of stems that have flowered.

Echinocactus polycephalus var. *polycephalus* (**cotton-top cactus**) grows on rocky slopes of the Mojave and northern Sonoran deserts at elevations below 1,000 m. Its large clusters of fat stems can be spotted at a distance. The short stems are about the size of a bowling ball (or human head), while the taller ones grow to about a half meter. The Greek *poly-* "many" plus *cephalus* "heads" refers to the clusters of stems looking somewhat like heads. The yellow flowers are not much appreciated, since they bloom in July or August and are almost overwhelmed by dense, stout spines.

Echinocereus (**hedgehog cactus**)—This genus has a common name matching the wrong scientific name—the result of numerous name changes and reclassifications. *Echinocereus* means "hedgehog" (*echino-*) "candle" (*cereus*), probably for the intense magenta or red flowers on the spiny plant. Large-lobed stigmas in the center of each flower are usually bright green. *Echinocereus* species have prominent ribs, either distinct or indistinct tubercles, and either straight or curved spines. Plants generally grow in low mounds formed either by low branching of stems or by clumping of stems tightly together. There are two easily differentiated species in California.

Echinocereus engelmannii (**strawberry hedgehog cactus, Engelmann's hedgehog cactus, calico cactus**) is named for the famous physician and botanist George Engelmann (1809–1884), who, with the resources of Henry Shaw, was instrumental in establishing the precursor of the Missouri Botanical Garden. Engelmann's hedgehog cactus grows in low clumps formed by the branching of numerous stems. Its flowers are usually intense magenta and bloom following spring rains. Young fruits are protected by spines, but as fruits mature they turn strawberry red and the spines fall away, inviting animals to indulge and thereby disperse seeds. Strawberry hedgehog cactus is widespread and fairly common in California's deserts at elevations below 2,400 m.

Echinocereus triglochidiatus (**Mojave mound cactus**) has bright orange-red spring-blooming flowers irresistible to hummingbirds. It forms dense mounds, sometimes composed of hundreds of spherical stems that are impenetrable both to competing vegetation and to herbivores. Mounds generally grow in rocky crevices or on slopes at elevations up to 3,000 m in both the Mojave and Sonoran Deserts. *Glochidiatus* means "provided with barbs," and *tri* means "three."

Escobaria (**beehive cactus, foxtail cactus**)—*Escobaria* is derived from the Spanish surname Escobar, while the common names for this genus cover the range of stem shapes, from a dome to a broad tail. *Escobaria* species are no more than 15 cm high and generally grow in clumps. Plants have inconspicuous ribs and prominent tubercles supporting dense, straight

Echinocereus engelmannii (strawberry hedgehog cactus)

Echinocereus triglochidiatus (Mojave mound cactus)

Escobaria vivipara var. *deserti* (beehive cactus)

Ferocactus cylindraceus (California barrel cactus)

spines. There are 16 species, all occurring in Mexico and the western United States. Only one species with three varieties grows in California's deserts.

Escobaria vivipara (**beehive cactus**) varieties are distinguishable by flower color, with the brighter ones rare and threatened with extinction by people seeking them out for collections and digging them out of their limited habitats. The formation of clumps by the sprouting of young stems at the base of a parent probably accounts for the specific epithet, meaning "sprouting" or "living" (*vivi-*) while "beside" (*-para*) the parent. All three varieties have the arrangement of spines characteristic of the genus: straight central spines extending out from the stem, surrounded by numerous radial spines spreading in a circle flat against the stem. The most common variety, *Escobaria vivipara* var. *deserti*, has delicate straw-colored flowers (sometimes yellow-green or pink) and grows on limestone soils in some eastern San Bernardino County mountain ranges at elevations of 1,000–2,400 m.

Ferocactus (**barrel cactus, visnaga**)— This genus is aptly named "fierce" (*fero-*) "cactus" for its dense, tough spines and its large, stout stature. The stature is also the basis for the name barrel cactus, although several species are taller and

skinnier than a typical barrel. All are single-stemmed, almost spherical when young, but soon adding height to become either barrel-like or columnar. Occasional specimens have reached a height of about 3 m. Barrel cacti are strongly ribbed, but tubercles are inconspicuous except for their dense spines. Of the 23 species in the genus, only two live in California. One of these, coast barrel cactus (*Ferocactus viridescens*), is uncommon on the coast of San Diego County, and the other is our desert species, *Ferocactus cylindraceus*.

Ferocactus cylindraceus (**California barrel cactus**) is one of the cylindrical, columnar members of the genus, generally growing to a height of 1–2 m. It has stout spines, both erect and spreading, the longest of which may be 17 cm. Spines are often red, especially on healthy young plants, and are usually curved, but not hooked. A stunning crown of yellow flowers forms on top of the column in spring. They are especially visible when viewed from the south, due to the sunward tilt of mature plants. By adopting this angle, barrel cacti cut down on the amount of surface area that is exposed to the most intense midday solar radiation, although an adaptive advantage of this has, to our knowledge, not been tested. California barrel cactus is uncommon, but certain parts of Anza Borrego State Park and the eastern Mojave Desert have large populations. It grows on rocky slopes and sandy or gravelly areas at elevations up to about 1,500 m.

Mammillaria (**nipple cactus, fish-hook cactus**)—*Mammil* means "nipple" in Latin, and indeed, this cactus is shaped a bit like nipples; however, the presence of spines makes for a bizarre comparison. Fish-hook cactus is descriptive of the recurved tips of longer spines of *Mammillaria*. Of the 150 species of fish-hook cactus, all in North America, only three grow in California.

Mammillaria dioica (**nipple cactus, fish-hook cactus**) generally has one to several stems and grows to a height of about 20 cm or occasionally 30 cm. Flowers are yellow to almost white, and some plants have flowers that are strictly female, with no stamens. *Dioica* means that the "two" (*di-*) sexes are in separate "houses" (*oeca*)— that is, on separate plants, a trait unusual for Cactaceae representatives. Cute little bright red fruits

Mammillaria dioica (nipple cactus)

from female flowers stick straight out from the stem, sometimes even remaining when a new round of flowers blooms. *Mammillaria dioica* is generally found in rocky washes of the western Sonoran Desert at elevations up to 1,500 m.

Opuntia (**prickly-pear, cholla**)—This is the largest genus in the Cactus Family, with about 200 species, several of which hybridize quite freely. *Opuntia* are probably native only to the Americas, although one species, Indian fig (*Opuntia ficus indica*), has been cultivated so widely for so long that its origins are unknown. Other species also have spread widely, either as weeds or as deliberate introductions, and now are common around the Mediterranean, in large parts of Africa and India, and throughout Australia.

   Opuntia are characterized by their jointed stems, which consist of pancakelike flat pads joined edge to edge, or sausagelike cylindrical tubes joined end to end, or almost any variation you can think of in almost any size. Most *Opuntia* have spines growing from areoles, like other cacti, and they all have glochids—small barbed bristles that are almost invisible. Fruits are generally fleshy, juicy, and edible, although picking them without getting glochids in your fingers can be tricky. Once

Opuntia basilaris var. *basilaris* (beavertail cactus), Anza Borrego Desert State Park, Hellhole Canyon

removed from the plant, fruits can be scorched briefly to burn off the glochids. These "prickles" on the "pears" are the basis for the common name. *Opun* is the Papago Indian name for the food plant and is probably the origin for the scientific name.

There are 21 species of *Opuntia* in California, including four of hybrid origin and one (Indian fig) that is nonnative. Only 12 of California's *Opuntia* species and two hybrids grow in the deserts. Most others are found near the southern coast.

***Opuntia basilaris* var. *basilaris* (beavertail cactus)** is unusual in that it lacks spines—though there are plenty of glochids growing at each areole. Stem segments are shaped like a beaver's tail and grow in clumps up to 40 cm high. Flowers are bright pink to magenta and often very abundant, growing along the upper edges of stem segments. The specific epithet *basilaris,* meaning "regal" in Greek, characterizes the beauty of plants in flower. Beavertail cactus is widespread, occurring throughout the deserts into adjacent mountains. One variety, found only in Kern County, is endangered, but the desert variety, *Opuntia basilaris* var. *basilaris,* is common.

Opuntia ramosissima (pencil cactus)

***Opuntia ramosissima* (pencil cactus, diamond cholla)** is named for its growth pattern—"most" (*-issima*) "branched" (*ramosa*)—and represents the thin-stemmed, highly branched end of the spectrum of *Opuntia* shapes. It is usually treelike and about 1 m tall, with long (3–4 cm) straight spines. Stem segments are only 1 cm or less in diameter and long like a pencil. Tubercles also are elongated in a diamond shape and prominent. Small orange-brown to pinkish flowers appear about June in rare years, so you have to get out of your air-conditioned car to appreciate them. Pencil cactus prefers sandy or gravelly desert flats at elevations under 1,100 m.

Sclerocactus (**pineapple cactus, devil-claw**)—*Sclero-,* the Greek word for "hard," could be used to describe most cacti. This generally single-stemmed genus is about the size and shape of a large pineapple, with both ribs and tubercles as well as numerous spines. Nineteen species inhabit Mexico and the southwestern United States, two of which grow in California's deserts.

Sclerocactus polyancistrus (**Mojave fishhook cactus**), with its "many" (*poly-*) "fishhooks" (*ancistrus*), is noted for its hooked spines, in contrast to the other California species, *Sclerocactus johnsonii*, which has straight or curved spines that are not hooked. *Sclerocactus polyancistrus* is uncommon, growing on limestone in hills and canyons of the Mojave Desert. Its large bright pink flowers attract attention in spring, and it is frequently the target of illegal collections.

Sclerocactus polyancistrus (Mojave fishhook cactus)

Examples and Uses

Examples of cacti can be found in most nurseries and botanical gardens, where they are raised from seed or grown from cuttings rather than removed from the wild. Through hybridization, serious collectors can develop new varieties from the hundreds of wild species found throughout the Western Hemisphere. Cacti are second only to orchids in popularity of collecting.

Landscaping with cactus is prevalent in arid climates and can cut both water use and maintenance time. Traditionally, cacti have been used as living fences, readily growing from cuttings. Most Cactaceae grow slowly, so the investment is for the long term.

Fleshy plant stems in an arid environment make cacti a target source of food and water for animals, including people. But how does the odd growth form of cacti function to their advantage in the desert? One unusual feature of cacti is the latticelike woody tissue of stems that allows them to expand and contract in response to the amount of water being stored. Another feature is the absence of leaves. Leaves can be a problem in desert environments because they have a lot of surface area through which the plant loses water. Many desert species have very small, tough leaves or leaves that fall off during drought, but this means that new leaves have to develop before a plant can benefit significantly from rare wet weather. Even when water is abundant, cacti depend entirely on their stems for photosynthesis. Stems exchange carbon dioxide with the atmosphere through small pores normally found mainly on a plant's leaves.

This leads to another interesting adaptation. Most plants have their pores open during the day to absorb atmospheric carbon dioxide, the carbon of which is then bound in the light-dependent process of photosynthesis into sugars for the plant's growth. Under desert conditions of intense heat and light, loss of water vapor through the open pores increases the risk of severe dehydration. Cacti have the ability to use an alternative form of photosynthesis when water availability is low, absorbing carbon dioxide at night (with pores open) and storing it for photosynthetic transformation into sugars during the day (with pores closed). This strategy of storage for later use ensures minimal loss of water, but at the cost of efficiency. However, when soil moisture is adequate for the plant to replace water that evaporates through open pores exposed to the sun, cacti can revert to the more efficient mode of photosynthesis.

There are stories about stranded pioneers extracting life-saving water from cactus stems, but most people would probably die from the wounds inflicted by spines or from alkaloids present in tissue. A few desert-dwelling Native Americans developed techniques for extracting and treating cactus flesh to get water or food, and some species of *Opuntia* with flat pancakelike stem joints still are eaten. Usually, immature pads with less oxalic acid are selected, cut into thin strips, cooked, and served like green beans. Both the plants and the edible pads are called *nopales*.

Many cactus fruits are edible, although only a few are tasty. Several species of *Opuntia* fruits (known as *tunas*) are sold in grocery stores in the Southwest, and some are harvested in the wild with little harm to the plants. A quick scorching or a thorough rolling in sand removes most prickles, and flesh can be scooped from split fruits. Fruits are eaten raw, squeezed for juice, cooked in soups and sauces, or made into jelly.

A few animals, including desert bighorn sheep, are able to eat some cactus stems, but the most useful parts of the plant are fruits and pollen and nectar of flowers. Sometimes cactus plants or skeletons of old plants are used by birds and rodents as nest sites, and cactus mounds make well-protected refuges for any small or burrowing species. However, the biggest rattlesnake we have seen in the desert was plundering rodent burrows around the base of *Opuntia parishii*—so it seems there are very few safe refuges.

CAPPARACEAE

CAPER FAMILY

Cleome lutea (yellow bee plant)

apers that we know from salads and sauces grow in arid Mediterranean areas and have been used in cooking for a very long time. We do not have the eating kinds of capers in our deserts, but if we did the flowers would be picked before they even had a chance to open. It is the flower bud, not the fruit, that tastes good.

Caper or capparis means "caper" in Greek, Latin, and old French and refers to the shrubs that provide capers as well as to other members of the family.

Capparaceae Icon Features

- Four free petals arranged with slightly bilateral symmetry
- Six stamens, long and curving
- Superior ovary, usually extended on a stalk

Distinctive Features of Capparaceae

Look for four free petals, usually yellow with slight bilateral symmetry, six long stamens, and an ovary on a stalk. If most of these features are combined with bad smelling leaves composed of three leaflets, you are probably looking at a Capparaceae member. Although these features are not consistent throughout the family, they are pretty reliable for species found in California's deserts. In a few cases you may find an ovary without a stalk, a flower that is unisexual (usually lacking the pistil), or one that is radial rather than bilateral in symmetry. Some plants may have a few leaves with more than three leaflets or upper leaves that are simple. Some species may even be blessed with an acceptable odor.

Stalked ovaries have the odd appearance of floating out in front of the flower, because the stalk is often very thin and about the same length as spreading filaments. Of course, fruits, which are mature ovaries with seeds, develop on these same thin stalks. Fruits are frequently oddly shaped, some with a persistent style that becomes spiny. Most species have four sepals that are either free or fused, and they often remain after the petals have fallen and the fruit has matured. Persistent sepals mark the junction between the flower's pedicel and the fruit's stalk.

Petal (4)

Stamen (6)

Style

Stalked ovary

Mojave stinkweed (*Cleomella obtusifolia*)

Caper Family (Capparaceae) flowers generally have four petals with bilateral symmetry, six long stamens, and an ovary on a stalk.

Similar Families

The Mustard Family (Brassicaceae) is right next to the Caper Family on the evolutionary tree, and some have suggested that the two families be combined. The most reliable difference between the families is the septum, or internal cross-wall, that divides fruits of Brassicaceae in half and that is lacking in Capparaceae. Technically, this is a small difference because the supporting structure for seeds, which also supports the septum, when present, is essentially the same in both families. Nevertheless, the presence or absence of a septum is easy to detect and quite a reliable distinction, with a few exceptions.

For California desert species, you do not even need fruits—just look at and smell the leaves to determine whether you have a Mustard or Caper Family representative. Brassicaceae rarely has anything resembling three leaflets, and mustard leaves may smell a bit spicy but not usually unpleasant.

Several other families, like the Pea and Bean Family (Fabaceae), have members with three leaflets, but their flowers do not have four similar, free petals in an almost radial arrangement.

Family Size and Distribution

Capparaceae comprises 45 genera and 800 species, mostly in tropical and subtropical zones throughout the world. The genus *Capparis* is the largest, with 350 species, none of which are found in California. Its most celebrated member, *Capparis spinosa*, native to southern Europe, is the source of capers at our dinner table and the namesake of the family.

Only six genera of Capparaceae are found in California, five of these in the desert. Half the California genera have only a single species, each of which occurs in our deserts. Most desert species of Capparaceae are found on strongly alkaline soils. In contrast to Brassicaceae, Capparaceae members are not very successful as weeds, and all California species are considered native.

California Desert Genera and Species

Most of our Caper Family members are annuals, but there is one perennial and one shrub, bladderpod (*Isomeris arborea*), which is widespread and notable for its inflated fruits. All desert species have flowers that are some shade of yellow, and all have six stamens, even though other genera commonly have more.

SIMPLIFIED KEY TO DESERT GENERA OF CAPPARACEAE

1. Shrub. Fruit inflated, with a lot of empty space inside .... *Isomeris arborea*
1' Annuals or rarely perennial. Fruits not mainly empty.
  2. Fruits elongated (15–45 mm) cylinders 5 mm or less in diameter (2 spp) .... *Cleome*
  2' Fruits generally wider than long, 2-lobed, often with an elongated style.
    3. Style in fruit elongated into a stiff sharp spine .... *Oxystylis lutea*
    3' Style in fruit flexible, elongated or not, but not a sharp spine.
      • Sepals free .... *Wislizenia refracta*
      • Sepals fused at the base (4 spp) .... *Cleomella*

Cleome (bee plant, spiderwort)—In early European languages, *Cleome* was the name for mustardlike plants. There are at least 150 species, and they have a preference for tropical and subtropical regions. Only four species occur in California, two of which are represented in our deserts. *Cleome* species are annuals, often tall and attractive. Flowers usually are bilateral and frequently are single sex with vestigial parts of the opposite sex present but nonfunctional.

Cleome sparsifolia (sparse-leaved bee plant) definitely is "sparse" (*sparsi-*) "leaved" (*folia*), as in none-at-all by flowering time. It is the intricate yellow-green branching that will attract your attention, and finally you will see some yellow flowers with petals about 1 cm in length. Fruits are 14–45 mm long and only 1–3 mm in diameter. Look for *Cleome sparsifolia* on sand dunes east of Owens Dry Lake or at other similar locations in the Mojave Desert.

Cleome sparsifolia (sparse-leaved bee plant)

Cleomella obtusifolia (Mojave stinkweed)

Cleomella (stinkweed)—This genus has a diminutive ending on the name *Cleome*; therefore, *Cleomella* means a "small mustardlike plant." *Cleomella* species may grow to nearly 1 m in height, but most of them are much smaller. All four of California's species of *Cleomella* are associated with alkaline soils of our deserts. Three species—*Cleomella brevipes*, *Cleomella parviflora*, and *Cleomella placasperma*—are specialized to moist habitats around thermal springs of the Mojave Desert. These three species have no hairs on stems or leaves.

Cleomella obtusifolia—(Mojave stinkweed, blunt-leaved stinkweed) is relatively widespread, occurring on sandy or rocky alkaline flats or disturbed roadsides, and it is the only hairy species of *Cleomella* in the Mojave Desert. The specific epithet refers to the fact that this species has "blunt" and "not so pointed" (*obtusi-*) "leaves" (*folia*) compared to other species. The common name is

Isomeris arborea (bladderpod)

self-explanatory. The stems of *Cleomella obtusi-folia* may be almost 1 m long, but these are the older ones that lie prostrate on the ground forming a large mat. Young stems in the center of the plant are upright. Fruits are only 3–4 mm, with a distinctive shape that looks almost like two tiny chocolate chips held base to base. This odd fruit grows on a thin, recurved stalk that is 6–8 mm in length. Opposite the stalk is the remaining small (1.5–5 mm) style of the flower's pistil.

Isomeris (bladderpod)—*Isomeris* means "equal" (*iso-*) "parts" (*meris*) in Greek and may refer to the nearly equal lengths of stamens and pistil. This genus has only one species, which is the only shrub of the family that occurs in California.

Isomeris arborea (bladderpod) can be a fairly large (to 2 m) shrub, as we may surmise from the specific epithet, *arborea*, meaning "treelike" in Latin. These shrubs flower profusely in the spring or after any good rains, and are an important source of nectar for hummingbirds. The yellow flowers are conspicuous, and later, the mature

Oxystylis lutea (false clover)

fruits inflate to a fat pod 3–4 cm long on a 1–2 cm stalk that cannot be missed. Occasionally flowers of *Isomeris arborea* lack their pistils, which will have aborted in the bud stage, so of course there will be no pods developing from these flowers. Bladderpod is common in washes and on sandy flats throughout the Mojave and Sonoran Deserts of California.

Oxystylis (**false clover**)—The common name refers to three bright green leaflets somewhat resembling clover, but distinctive fruits with a "sharp" (*oxy-*) "style" (*stylis*) reveal the true identity of this single-species genus.

Oxystylis lutea (**false clover**) bears the specific epithet of *lutea* (yellowish) because the flowers are yellow and even the leaves and stem are yellowish green. It is an annual but can grow to over 1 m tall, standing out distinctly against a dry desert background. Flowers grow in tight clusters in leaf axils along thick stems. Petals are only 2–3 mm, so you could miss the flowers entirely if you failed to look inside the leafy exterior. *Oxystylis lutea* likes alkaline flats and roadsides of the Mojave Desert.

Examples and Uses

Capers in the grocery store come in small, expensive glass bottles. They are the pickled flower buds of *Capparis spinosa*, a prickly crawly shrub native to southern Europe. Capers have been used in the Mediterranean region for over 2,000 years, but they are virtually the only food-bearing member of the family. Spain, Italy, and Algeria still produce most of the capers distributed around the world.

Cucurbita palmata (coyote melon)

CUCURBITACEAE

GOURD OR CUCUMBER FAMILY

What a surprise it can be to find large, leafy Gourd Family vines in the desert, or the fruits, which are often as big as an orange. Although Gourd Family members have been cultivated around the globe for thousands of years, resulting in watermelons and the like, in our deserts the fruits do not make much of a meal, nor should they be relied upon for emergency water. They are big and beautiful, but dry.

Cucurbita, Latin for "gourd," is the basis for the family name Cucurbitaceae, while the Greek word for cucumber, *Cucumis* (probably from the same root), gives us the genus name of that fruit.

Cucurbitaceae Icon Features

- Vines with tendrils for climbing
- Alternate leaves often palmately lobed and veined
- Sexes in separate flowers

Distinctive Features of Cucurbitaceae

Look for vines, large or small, on the ground or in shrubs or trees, with tendrils and alternate leaves. Stems, if cut in cross section, often have five ridges or angles. Stems may be fat and tough, but they are never woody. Leaves are usually palmately lobed. The single-sex flowers grow in the axils of leaves or at branching points of the stem, or both, and are radially symmetrical. The sexes may occur on the same plant or on separate plants. Usually several male flowers grow on a short stem from a leaf axil or branching point, whereas female flowers tend to be solitary at leaf axils or branching points.

The ovary is inferior, a tiny version of the fruit easily observed below where petals are attached on female flowers. Petals are fused together into a saucer or cup shape with five lobes. The style, usually with three large stigma lobes, is inside the cup-shaped corolla of the female flower. This simple unisex arrangement can sometimes appear ambiguous when there are also a few infertile stamens without anthers. Male flowers have five stamens inside their own corolla, but the stamens are sometimes joined in various ways to appear fewer in number, resembling female flowers. Both stigma lobes and anthers seem quite oversized and contorted in this family, causing them to look sufficiently similar that pollen-collecting insects are attracted into both sexes of flowers and accidentally visit females. The insects then leave a bit of pollen on the stigmas and so effect cross-fertilization.

Most Cucurbitaceae fruits look like some version of a cucumber, squash, pumpkin, or watermelon, with numerous large flat seeds inside watery, fleshy tissue surrounded by rind. However, several species have only a single seed or a small number of them. Some genera have spiny fruits, unfamiliar to us from our culinary experience with the family.

Similar Families

There is not much that can be confused with Cucurbitaceae in California's deserts. Simply insist on single-sex flowers on a vine with alternate leaves. There are not many vines in the desert, and the others present have opposite leaves, woody stems, or other obvious differences such that they do not resemble Cucurbitaceae members.

Family Size and Distribution

Cucurbitaceae has around 100 genera and at least 700 species. Although weighted toward the tropics, the family is found worldwide, and some important food species, such as squashes and pumpkins, originated in arid regions of the Americas. Cucumber has been cultivated for so long that it is unclear whether it originated in India or Africa.

There are only six genera of Cucurbitaceae in California (not counting grocery stores), and three of them are nonnatives. Four genera are represented in our deserts, all native except *Cucumis*, or cucumber, which is well established as a weed in some areas, including the southeastern Sonoran Desert in Imperial County. These weeds are classified as "noxious" because of their aggressive habits and the difficulty of removing them.

Most of our desert species of Cucurbitaceae are found in areas with a bit of extra moisture, such as roadsides and washes or edges of the desert bordering more moist ecosystems. Look upward to find the smaller-leaved genera, because they are prolific climbers, often covering shrubs and small trees.

California Desert Genera and Species

All desert species of Cucurbitaceae in California grow as perennial vines, usually with a large underground tuber, or expanded root, similar to a huge potato. This underground storage organ allows these species to sprout quickly following seasonal droughts or disturbances such as fire. Vines can then spread rapidly and develop broad leaves, covering slower-growing plants and taking advantage of short seasons of favorable conditions. After development of fruits, leaves and stems dry up, leaving their accumulated nutrients either in the seeds and fruits or in the storage root to wait for the next short season. When you see Cucurbitaceae plants in the desert in early spring, they often seem out of place because of their lush, almost-tropical appearance. Go back a few months later and you will find only the dried skeletons of vines and a few orange-sized fruits.

SIMPLIFIED KEY TO DESERT GENERA OF CUCURBITACEAE

1. Corolla white, cream, or greenish.
  • Fruit body less than 1 cm long .... *Brandegea bigelovii*
  • Fruit body 4 cm or more long (2 spp) .... *Marah*
1' Corolla yellow to orange-yellow.
  • Corolla 2–3 cm wide, base fused for 1 cm. Tendrils unbranched .... *Cucumis melo*
  • Corolla more than 3 cm wide, base fused for 4 cm or more. Tendrils generally branched (3 spp) .... *Cucurbita*

Brandegea—This genus is named after the California botanist T. S. Brandegee (1843–1925), who, with his wife, contributed many specimens and much experience to the University of California Herbarium. The single species has the smallest flower and fruit of the Cucurbitaceae in California's deserts.

Brandegea bigelovii (Brandegea) does not have anything to do with "big"; rather, *bigelovii* refers to Dr. Jacob M. Bigelow (1787–1879), author of *American Medical Botany* and a professor at Harvard Medical School who collected hundreds of plants in California during the 1850s. Although this species, named after two botanists, is true to the classic features of its family, it requires a close look because of its diminutive size. Flowers are only 1.5–3 mm wide, and fruits only 5–6 mm long, not including a beak (the dried style and stigma) of similar length. Leaves of *Brandegea bigelovii* are highly variable versions of the palmate form: some are so deeply lobed between the main veins that they resemble a hand that is all fingers and no palm, while others have

Brandegea bigelovii (brandegea)

very rounded lobes or very short lobes, almost seeming to come from a different plant. *Brandegea bigelovii* grows from a substantial taproot but does not have underground tubers. It is frequently found in canyons and washes at lower elevations throughout California's deserts.

Cucurbita—This genus has about 30 species, all native to the Americas. Three species occur in California, all in deserts and other low-elevation arid habitats. All have large underground tubers that store fluids and nutrients, enabling new vines and roots to grow rapidly following an extended seasonal drought. Thanks to these emergency supplies, the plant's leaves are not required to be quite so drought tolerant as those of other desert species. Thus, the rather large (up to 15 or even 30 cm) spreading leaves of *Cucurbita* species contrast strongly with the small stiff leaves of the desert shrubs growing around them. When soil moisture is exhausted, leaves simply wither and die, to be replaced the following year. Fruits in this genus are similar to some of the small gourds sold for decoration in the fall—round and sometimes striped green, white, or yellow when mature. They may be quite colorful and obvious among withered leaves of a spent vine, or they may be mostly

Cucurbita palmata (coyote melon)

green and almost hidden beneath green leaves of a vigorous vine. Each species has yellow or yellow-orange flowers, 3 cm or more wide. Leaves are the most obvious way to distinguish among the species.

Cucurbita palmata (**coyote melon**) has "palmately" (*palmata*) divided leaves with five wide triangular lobes rather than slim lobes as in *Cucurbita digitata*. Leaves are 8–15 cm in length, and round fruits are 8–9 cm in diameter. The third species, *Cucurbita foetidissima*, has larger somewhat triangular leaves with a disgusting odor. Coyote melons grow in sandy soils at elevations under 1,300 m in both deserts.

Marah (**man-root, wild cucumber**)—The term *marah* means "bitter" in Latin, and is used as the name for bitter waters in the Bible. Every part of the plant is bitter and inedible, even the giant root that sometimes is as large as a man—hence the common name man-root. However, it is not necessary to taste this plant or to dig it up to identify it. All species of *Marah* have whitish (or yellowish-green) flowers and prickly fruits with relatively few seeds. There are seven species in western North America, two of which live in California's deserts.

Marah macrocarpus var. *macrocarpus* (**large-fruited man-root, wild cucumber, chilicothe**), as the name implies, is "large" (*macro-*) "fruited" (*carpus*), up to 12 cm long; it is also oblong in shape and seriously prickly. There are generally 4–12 roundish, shiny, dark seeds, which native people used as beads. Because of the huge tuber, *Marah macrocarpus* is especially well adapted to sprouting quickly after drought or fire, and it grows rapidly across shrubs or trees in its path, forming a bright green layer of broad five- to seven-lobed leaves about 10 cm long. You can't miss it! Look for several small (8–13 mm wide) male flowers on short stalks in leaf axils. The solitary and slightly larger female flower has a tiny fruit-form ovary below the petals and is usually positioned just at the base of the group of male flowers. Early in the season there may be only male flowers, but it will be worth a return visit to see the fruits. *Marah macrocarpus* grows in the Sonoran Desert, often in washes or transition zones bordering chaparral.

Examples and Uses

Probably everyone has a few favorite foods in Cucurbitaceae. The genus *Cucumis*, which provides California with two noxious weeds, also produces honeydew melon and cantaloupe. The genus *Citrullus* includes watermelon, originally from southern Africa, but cultivated in Egypt 6,000 years ago.

All of the various squashes, gourds, and pumpkins are from the Western Hemisphere and were introduced into the world diet via native people who were visited by European explorers starting in 1492. These edible fruits are all of the genus *Cucurbita*, and some have been in cultivation for over 9,000 years. Varieties we call summer squash, such as zucchini and crookneck, are immature fruits, as you can tell from the soft, poorly developed seeds. Winter squashes such as butternut and acorn are mature fruits with a hard outer shell, a starchy interior, and seeds with a hardened coat. Chayote (*Sechium*) is eaten like squash but has a single large seed rather than numerous flat ones.

Marah macrocarpus var. *macrocarpus* (large-fruited man-root)

In spite of the family's reputation for supplying important foods, desert species in California have little nutritional value. Neither the large watery tubers nor the fruits are edible. However, seeds of some species were consumed by desert people, and shells of *Cucurbita* fruits have served as containers.

Loofa sponges, popular in the bath, are veins of fruits of the genus *Luffa*. Soft tissue shrinks away, leaving reinforced veins that dry in a spongelike lattice formation. Sometimes fruits of *Marah macrocarpus* are used in a similar manner.

Cercidium floridum ssp. *floridum* (blue palo verde)

FABACEAE (LEGUMINOSAE)

LEGUME OR BEAN OR PEA FAMILY

Fruits and seeds of Fabaceae species provide food the world over, and were essential to early inhabitants of California's deserts. Fabaceae flowers are among the most abundant and spectacular of our deserts, attracting both human and insect visitors. The family is huge and diverse, taking forms from tiny annuals to trees, but all Fabaceae species have one thing in common: the fruit.

The word "legume" comes from the Latin *legumen,* meaning a fruit like a bean. Fabaceae replaces the older family name, Leguminosae. The Latin word *fabula,* meaning "story," was transformed in Old French into "fable." Think of the fable of Jack and the Bean Stalk, and you've conjoined *fab* and beans for permanent memory banking. More mundanely, *faba* or *fava* is the name for the Old World bean known to the Romans, whose prominent families, such as Fabius, were named after beans.

Fabaceae Icon Features

- Superior ovary forming pea or beanlike fruits
- Two to several seeds, usually large and spherical to ovoid
- Fruits generally splitting in halves with adjacent seeds attached to opposite halves

Distinctive Features of the Family

If it looks like a bean, it's a bean. With some exceptions, key features (slightly simplified) of the fruit known as a **legume** are that it has a single cavity that splits open along two sides, and it generally has two or more seeds attached on alternate sides of only one of the lines (sutures) where the fruit will split open. Thus, although usually seeds are in a single row, like peas in a pod, if you pull the pod apart along both sutures, adjacent peas will separate, remaining attached to opposite sides of the pod.

Fabaceae ovaries are superior, one per flower, and the legumes they grow into as fruits stick out in front of fading or fallen petals. Legumes are usually elongated and somewhat flattened, with the sutures forming the edges—looking much like the garden-variety pea or bean—but numerous variations exist, such as legumes that grow into a spiral shape (*Prosopis pubescens,* screw bean mesquite) or that have prickly pods (*Glycyrrhiza lepidota,* wild licorice).

Flowers, rather than fruits, are the basis for distinguishing the three main, taxonomically coherent subgroups of Fabaceae. (Some taxonomic references consider these subfamilies, consisting of closely related species, to be stand-alone families.) The most obvious difference among subfamilies lies in flower symmetry, which is radial, almost bilateral, or clearly bilateral. Other differences include the arrangement and position of stamens and whether or how petals are attached to one another.

Mimosoideae flowers have radial symmetry and grow in large clusters. The five tiny petals usually are joined into a tube with five equal lobes, but this arrangement is dominated by long stamens extending well beyond the corolla. Stamens are either entirely separate or fused at the base. There may be only 10 stamens, but often there are many, so clustered flowers resemble a pompom.

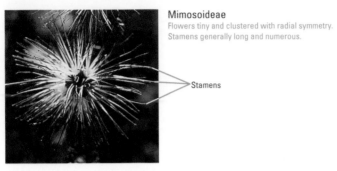

Mimosoideae
Flowers tiny and clustered with radial symmetry.
Stamens generally long and numerous.

Stamens

Fairyduster (*Calliandra eriophylla*)

Caesalpinioideae
Flowers larger and single or in small clusters.
Symmetry is not-quite-radial, or almost-bilateral.
Stamens visible but not dominant over petals.

Stamen
Petal

Spiny senna (*Senna armata*)

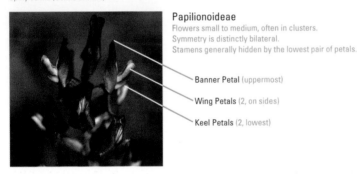

Papilionoideae
Flowers small to medium, often in clusters.
Symmetry is distinctly bilateral.
Stamens generally hidden by the lowest pair of petals.

Banner Petal (uppermost)
Wing Petals (2, on sides)
Keel Petals (2, lowest)

Freckled milkvetch (*Astragalus lentiginosus*
var. *fremontii*)

The Pea and Bean Family (Fabaceae) has three major subgroups characterized mainly by different types of floral symmetry and different features of the stamens.

Caesalpinioideae flowers are the in-between group, having not-quite-radial symmetry or almost-bilateral symmetry. There are five separate petals, not entirely equal in size and not entirely radial in position, but usually conspicuous. The 10 or fewer stamens are all separate, of the same general length as the petals, and visible among the petals.

Papilionoideae, by far the largest subgroup, has flowers that are clearly bilaterally symmetrical, with the 10 stamens generally hidden in the petals. Stamens usually are arranged as nine fused together plus one separate, although the 10 may be all separate or all fused.

Fabaceae includes annuals, perennials, shrubs, and trees. However, in California's deserts most Mimosoideae and Caesalpinioideae representatives are shrubs and trees, while most Papilionoideae members are annuals and perennials—but with notable exceptions.

Nearly all Fabaceae members have compound leaves. Leaves, in an alternate arrangement along stems, have stipules in the form of bumps, flags, or spines. Sometimes stipules are quite large and conspicuous, and they can help to distinguish a leaf from a leaflet because leaflets do not have stipules.

Similar Families

Superficially, fruits of desert willow, *Chilopsis linearis*, look like long beans. On closer examination, desert willow pods have two separate chambers, and seeds are winged and attached along opposite sides of each chamber. These are not legumes. Although flowers of desert willow have bilateral symmetry like many Fabaceae flowers, the petals, rather than being separate from one another, are fused into a large tube with unequal lobes. Finally, desert willow has long narrow leaves, similar to a willow's but very unlike any of our desert Fabaceae members. Desert willow is the only member of Bignoniaceae native to California.

Family Size and Distribution

There are roughly 650 genera and 18,000 species of Fabaceae worldwide. Many species are tropical or subtropical, but temperate climate representatives are abundant. Members of the family are frequently cultivated, and crop plants originating on different continents have been spread worldwide.

One of the largest genera, with over 2,000 species, is *Astragalus*, or locoweed, which is common in California's deserts. The genus *Acacia* has about 1,200 species, many of which are found in arid parts of Africa and Australia, but there is one California native, and it is a desert species. *Acacia* species are widely cultivated, and although many of those brought to California have spread from landscaping into surrounding areas, most are unable to naturalize without more water than our deserts provide.

California has 50 genera of Fabaceae, 19 of which comprise only nonnative species. The remaining 31 genera account for all of our native species as well as a liberal scattering of nonnatives. The California deserts have 24 genera of Fabaceae, only two of which are exclusively nonnative. Not surprisingly, *Astragalus* is the largest genus in our deserts, with 38 species (and many more varieties), all native to California.

Shrubs and trees of the family are more common at low elevations than high, and many are very deeply rooted through sandy or gravelly soils, reaching to groundwater or dependent on summer as well as winter rains and flash floods. These species may be found near springs, around margins of dry lakes, and along stream or flood courses that usually are dry.

California Desert Genera and Species

Blue and yellow are the most common colors in desert Fabaceae flowers, with white and rose to magenta much less common. Many members of the family are shrubs or trees; perennials are common, and annuals are uncommon, at least in our deserts. Compound leaves and fruits like a bean are the best general indicators.

SIMPLIFIED KEYS TO DESERT GENERA OF FABACEAE

The three subfamilies of Fabaceae have very different looks and are used below as the first distinction in determining genus. A separate key is provided for each subfamily of Fabaceae. Photographs and text that follow the keys also are arranged by subfamily.

KEY TO SUBFAMILY GROUP

1. Flowers with radial symmetry. Corollas tiny and usually tubular. Stamens much longer than corolla tubes and generally numerous (sometimes 10) .... **GROUP 1: Mimosoideae**

1' Flowers with bilateral symmetry, either subtle or distinct. Petals mostly separate and conspicuous. Stamens 10 or fewer, similar in length to petals.

2. Flowers with subtle bilateral symmetry. Petals similar in shape, sometimes differing in color. Stamens clearly visible .... GROUP 2: Caesalpinioideae

2' Flowers with distinct bilateral symmetry. Petals with different specialized shapes. Stamens generally hidden by 2 lower petals .... GROUP 3: Papilionoideae

KEYS TO GENERA BY SUBFAMILY GROUP

Various versions of compound leaves are true to genera and are used extensively in keys. **Once-compound** leaves have leaflets that are not divided. Once-compound leaves that are **palmate** have leaflets spreading from a single point like fingers from your palm. **Pinnate** arrangements of leaflets have many varieties, but the common feature is leaflets arranged along a central axis, like a "feather" or *pinna*. Leaves are **odd-pinnate** if there is an odd number of leaflets, or **even-pinnate** if there is an even number. Rather than counting tiny leaflets, just look to see if there is a leaflet at the tip. If not, the leaf only has pairs of leaflets and is even-pinnate, even if pairing is imperfect. If present, the leaflet at the tip makes for an odd-pinnate leaf. **Twice-compound** leaves have primary leaflets each of which is divided into secondary leaflets. Either set of leaflets may be even or odd.

Glands that look like dots or scale insect infestations on the surface of stems and leaves are another feature important in keys. The term **gland-dotted** refers to these surface structures but not to glandular hairs or smaller glands that are not distinctly dots.

When determining features of leaflets, be sure to look at many leaves from different parts of the plant. Factors such as maturity, sun exposure, and herbivory affect leaf size and shape as well as density of glands, so it is important to notice the range of possibilities before determining what is representative.

GROUP 1 KEY (MIMOSOIDEAE): Flowers with radial symmetry.

1. Plants unarmed. Flowers pink to reddish, mainly from coloring of stamens .... *Calliandra eriophylla*

1' Plants armed with spines or prickles. Flowers cream to yellow.

2. Stems with prickles curved like a cat's claw. Stamens many .... *Acacia greggii*

2' Stems with pairs of straight spines. Stamens 10 (3 spp) .... *Prosopis*

GROUP 2 KEY (CAESALPINIOIDEAE): Flowers with subtle bilateral symmetry.

1. Leaves once-pinnate, 2–8 pairs of leaflets (2 spp) .... *Senna*

1' Leaves twice-pinnate.

2. Trees with green trunks and branches. Thorns in some leaf axils. Leaves with 1 pair of primary leaflets and 2–8 pairs of secondary leaflets (2 spp) .... *Cercidium*

2' Shrubs or subshrubs, 1 m or less in height. Unarmed. Branches green or not. Leaves with 3 or more primary leaflets and 5–10 pairs of secondary leaflets.

• Shrubs with long arching green branches. Hairy, not glandular. Leaves with 3 primary

leaflets, center one longest with 8–10 pairs of secondary leaflets .... *Caesalpinia virgata*
• Subshrubs under 30 cm, glandular all over. Leaves with 5–11 primary leaflets, each with 5–10 pairs of secondary leaflets .... *Hoffmannseggia glauca*

GROUP 3 KEY (PAPILIONOIDEAE): Flowers with distinct bilateral symmetry.

1. Leaves simple.
 • Leaves oval to 2 cm. Plants without glands ....*Alhagi pseudalhagi*
 • Leaves linear to 3 cm. Plants gland-dotted (2 spp) ....**Psorothamnus**
1' Leaves compound.
  2. Leaves palmate or leaflets 3 in number (either palmate or pinnate).
  3. Leaflets 5 or more (palmate).
   • Leaflets gland-dotted. Fruits with 1 seed .... *Pediomelum castoreum*
   • Leaflets not gland-dotted. Fruits with 2 or more seeds (14 spp) .... **Lupinus**
  3' Leaflets generally 3 (some leaves on the same plant occasionally with more).
   4. Corolla pink or pink-purple.
    • Perennial, vinelike. Leaflets roughly triangular in outline, sometimes with 3 broadly rounded lobes .... *Phaseolus filiformis*
    • Annual, not vinelike. Leaflets oval to round, tips wider and notched (2 spp) .... *Trifolium*
   4' Corolla yellow, white, or purple with white tips
    • Corolla yellow. Leaflets generally 3, some 4 or 5, irregularly arranged, some appearing palmate, some pinnate, never toothed (3 spp) .... *Lotus*
    Corolla yellow or white. Leaflets 3, toothed. Plants erect (3 spp) ....*Melilotus*
    • Corolla white to purple with white tips. Plants matted .... *Astragalus calycosis*
  2' Leaflets once-pinnate, most or all leaves with more than 3 leaflets.
   5. Leaves even-once-pinnate.
    • Tree. Corolla at least partly purple .... *Olneya tesota*
    • Perennial, vinelike with tendrils. Corolla lilac to purple .... *Lathyrus hitchcockianus*
    • Annual. Corolla yellowish or spotted .... *Sesbania exaltata*
   5' Leaves odd-once-pinnate.
    6. Tree, shrub, or subshrub.
     • Corolla pink or white, 2–2.5 cm .... *Robinia neomexicana*
     • Corolla blue purple, sometimes with white, 1 cm or less (4 spp) ....**Psorothamnus**
    6' Annual or perennial.
     7. Perennial with stipules that are spines. Corolla pink or white .... *Peteria thompsoniae*
     7' Plants without spinelike stipules. Corollas variously colored.
      8. Fruits prickly. Corolla yellowish or greenish white .... *Glycyrrhiza lepidota*
      8' Fruits not prickly. Corollas variously colored.
       9. Corolla yellow. Leaflets generally 4–6, irregularly arranged; occasionally some leaves with up to 11 leaflets, or as few as 3 (10 spp) .... *Lotus*

9' Corolla not yellow, some cream. Leaflets generally at least 7.
  • Plants weakly gland-dotted. Stem long and weak, often trailing across shrubs. Leaves sparse, leaflets 11–23, neatly paired along axis. Corolla purple with white ....
  .... *Marina parryi*
  • Plants heavily gland dotted. Stems either short and spread on the ground or 30–50 cm and spreading above ground. Leaflets 5–12. Corolla whitish lavender or purple (3 spp) .... *Dalea*
  • Plants not gland-dotted. Stems, short or medium, leafy. Leaflets usually at least 7, up to 33. Corollas rose, lavender, or purple, some white to cream, rarely red (37 spp) ....
  .... *Astragalus*

California Desert Mimosoideae All three genera of California desert Mimosoideae are shrubs or small trees, recognizable by their tight clusters of tiny flowers with extra-long stamens, a strong fragrance, and a large retinue of attending bees. Leaves are twice-compound. Primary leaflets are usually in one, two, or three pairs; secondary leaflets are usually in several to many pairs.

Acacia (acacia)—The Greek word *akakie,* meaning "sharp point," describes the well-armed status of many members of the genus. California has 13 naturalized species (mainly Australian) and only one native. The native is the only species found in California's deserts outside of cultivation.

Acacia greggii (catclaw acacia) is named after Josiah Gregg (1806–1850), who explored and wrote about the arid Southwest. In Mexico he collected plants and sent specimens to experts, hoping to find material new to botanists. The common name, catclaw acacia, describes the shape and sharpness of prickles growing along the woody stems of the plant. There are two or three pairs of primary leaflets and fewer than 10 pairs of secondary leaflets. Tiny pale yellow flowers with many stamens appear in spikes clustered along the leafy ends of stems. Fruits are recognizable as beans, 5–15 cm long, but they may be twisted, somewhat flattened, and shrunken between the seeds. *Acacia greggii* is found

Acacia greggii (catclaw)

on flats and in washes at elevations below about 1,400 m.

Calliandra—The combination of Greek words *calli-*, meaning "beautiful," and -*andra*, referring to the "male" parts, gives us "beautiful stamens," which are, indeed, the focal point of the flower. Of some 200 species of *Calliandra*, only one is found in California.

Calliandra eriophylla (fairyduster) has such long (2 cm), bright pink, lovely stamens that it is easy to imagine fairies using the flowers when we are not looking. However,

Calliandra eriophylla (fairyduster)

the specific epithet, which is Greek, refers to the "woolly" (*erio-*) "leaves" (*phylla*). Finely hairy is a better description for the leaves, which otherwise are built a lot like catclaw acacia leaves, with two to four pairs of primary leaflets and seven to nine pairs of secondary leaflets. Fruits are about 5 cm long, flat, and hairy, but generally beanlike. Fairyduster is a small unarmed shrub, confined, in California, to the Sonoran Desert.

Prosopis (**mesquite**)—*Prosopis* is the Greek name for a certain type of bur in the Sunflower Family, bearing little or no resemblance to mesquite. Mesquite is a Mexican word, adapted from an Indian name for these small trees. There are two native species of *Prosopis* in California's deserts, plus one species that is naturalized in a very limited portion of Imperial County.

Mesquite is an excellent indicator of water, growing at springs, seeps, and where there is relatively shallow groundwater. Thus, mesquite was both helpful in finding water and available to settlers near water. Indians and early settlers utilized wood for fires and structures, and beans and seeds for food for themselves and their livestock. Use of mesquite wood for grilling has been adopted in the modern world of fine restaurants and backyard barbecues.

The tiny flowers of *Prosopis,* which hang in large catkinlike clusters with the twice-compound leaves along the ends of branches, are radial in symmetry, but with just 10 stamens rather than the numerous ones present in other California desert Mimosoideae.

Prosopis species are well armed, their stipules forming stiff spines, one on each side of the attachment of leaf to stem.

Prosopis glandulosa var. *torreyana* (**honey mesquite**) is the more common of the two native species of *Prosopis*. Its specific epithet, *glandulosa*, meaning that it is glandular, is somewhat misleading, since most of the plant is hairless and glandless; however, most mesquites have small sunken glands between paired leaflets, and anthers have tiny glands on their tips. The varietal name honors John Torrey (1796–1873), a prominent New York plant taxonomist who classified many of the specimens shipped east by various explorers of new western states and territories. This is the only variety of *Prosopis glandulosa* in California. Honey mesquite undoubtedly is a name that honors the excellent honey that bees make from the abundant flowers.

Honey mesquite usually has only one pair of primary leaflets but there are numerous secondary leaflets, which are up to 2.5 cm long, narrow, and oblong. The flowers are tiny and cream or greenish, but they break the rules by having five separate petals rather than fused petals, which are characteristic of Mimosoideae. Don't worry, though, because the petals are so small that you will know this is

Prosopis glandulosa var. *torreyana* (honey mesquite)

Prosopis pubescens (screw bean mesquite)

mesquite before you have a chance to inspect them. Fruits are typical beans. Honey mesquite is common at lower elevations (under 1,700 m) throughout the deserts, wherever their long roots can reach water. Sometimes they even grow in sand dunes, if they have been able to get established where there was some accumulation of water.

Prosopis pubescens (**screw bean mesquite, tornillo**) is rather uncommon, and its tightly coiled beans, looking much like compressed corkscrews—as the Spanish word *tornillo*, meaning "screw," suggests—are unmistakable. *Pubescens* comes from Latin, meaning "having small downy hairs"—in this case, on the leaves. There are one or two pairs of primary leaflets and five to eight pairs of secondary leaflets, shorter than those of honey mesquite. In conformance with its subfamily characteristics, the tiny petals of screw bean mesquite are fused. *Prosopis pubescens* is found in washes, creek beds, and river bottoms at elevations under 1,300 m.

California Desert Caesalpinioideae Four genera of Caesalpinioideae inhabit the California deserts, most of which are shrubs or small trees, with one subshrub. Three genera are described below and the fourth, *Hoffmannseggia*, is an uncommon but aggressive weed. Flowers of Caesalpinioideae are generally yellow, with conspicuous separate petals not quite in radial symmetry, and 10 stamens. Leaves are once- or twice-compound. All fruits look pretty much like beans or peas.

Caesalpinia—A single native species of *Caesalpinia* is found in California, a rather uncommon inhabitant of the Sonoran Desert. Overall, the genus, named after Andreas Caesalpini (1519–1603), an Italian botanist and physician to Pope Clement VIII, has about 200 species, mainly in tropical and warm-temperate regions.

Caesalpinia virgata (**littleleaf rushpea**) is a shrub with long slender green twigs (or, in Latin, *virgata*). The branches are hairy and usually leafless, having a wand- or rushlike appearance. The twice-compound leaves, growing in response to early spring rains, have three primary leaflets, with the middle one longer than the other two. There are three to six pairs of secondary leaflets on the shorter primaries, and eight to 10 pairs on the longer central primary leaflet. Flowers tend to be sparse. The petals are 6–8 mm long and bright yellow with some reddish markings, and the uppermost petal generally fades to red. The 10 stamens also are yellow and about the same length as petals. Fruits, glandular when young, look like small pea pods.

Cercidium (**palo verde**)—There are two species of *Cercidium* in California, both natives and desert inhabitants. They are small, heavily branched, and well-armed trees, with bark that is noticeably green, even on older wood. This accounts for the Spanish name meaning "green" (*verde*) "stick" (*palo*). *Cercidium*, the Greek word for a weaver's shuttle, refers to the shape of the fruit— basically, a flattened bean. Leaves are twice-compound, with only one pair of primary leaflets. Flowers are about 1 cm across, with separate petals, slightly bilateral symmetry, and 10 conspicuous stamens.

Cercidium floridum **ssp.** *floridum* (**blue palo verde**) is spectacularly "flowery" (*floridum*) in spring, when trees are entirely cloaked in bright yellow. The term "blue" in the common name refers to a somewhat bluish tint to the green of leaves and stems, both of which are without hairs. Each of the primary leaflets on blue palo verde has only two or three pairs of secondary leaflets, but leaves may be hard to find most of the year. This *Cercidium* grows in flood plains and washes at elevations around 1,100 m, mostly in the Sonoran Desert.

Caesalpinia virgata (littleleaf rushpea)

Cercidium floridum ssp. *floridum* (blue palo verde)

Senna (**cassia**)—This genus (until recently called *Cassia* in North America) has about 260 members overall, with only two species native to California, both of them found in our deserts. *Senna* comes from the Arabic *sana*, a general term for cassia species present in that part of the world. Leaves of *Senna* are once-compound, and the California native species have only two to four pairs of leaflets. Flower petals are yellow and showy, up to 12 mm long, and arranged with slight bilateral symmetry. There are 10 stamens, but only seven of them are fertile. Anthers have tiny holes at their tips where pollen is released in response to vibrations caused by visiting bees. Fruits are typical beans.

Senna covesii (**Coues' cassia**) is an unarmed subshrub about a half meter or less in height and almost furry with white hairs. Leaves have only two to three pairs of rounded leaflets. Petals also are almost round and about 1 cm long, in an almost radial arrangement. The specific epithet *covesii* honors the naturalist Dr. Elliott Coues (1842–1899), who collected mainly in Arizona and who wrote *Birds of the Colorado Valley. Senna covesii* grows in sandy habitats of the Sonoran Desert from 500 to 600 m.

California Desert Papilionoideae All remaining California desert genera of Fabaceae are in the subfamily Papilionoideae, including the huge genus *Astragalus* and the well-known *Lupinus*. There are annuals, lots of perennials, some shrubs, and a few trees. Purples and blues or magentas are common flower colors, but there are yellows, creams, and whites as well, and even red. Leaves are once-compound, either with the usual pinnate arrangement of leaflets or with a palmate arrangement.

The Papilionoideae flower is designed to enlist bees in the business of cross-pollination. The uppermost petal, called the **banner**, bends up and back above the others as an attractant (banner) signaling pollinators. Below the banner, one petal on each side, called **wings**, serve as a landing platform for incoming aviators. At the bottom, two petals cupped together and slightly attached form the **keel**. The keel encloses the stamens and pistil, exposing their tips (anthers and stigma) in response to pressure on the landing platform above. Thus, a bee, seeing the banner, lands on the wings to get nectar from near the base of the ovary. The bee's weight causes the anthers and stigma to extend out of the keel, exposing the stigma to the bee's underbelly, which is carrying pollen from a visit to a previous flower. The exposed anthers brush new pollen into place, which the bee then carries to another flower where it is likely to contact another stigma.

Senna covesii (Coues' cassia)

Astragalus (**locoweed, milkvetch**)—The Greek word *Astragalus* refers to a curved bone in the human foot that is shaped like the inflated pods of many members of this genus. The common name locoweed refers to the condition of cattle that have grazed on *Astragalus*. California has 94 of the more than 2,000 species worldwide, 38 of which are found in California's deserts. Most are extremely difficult to identify, and many have numerous varieties that are even harder to pin down. Both fruits and flowers are needed for most identifications. Features like hair structure, the amount of reflexing of the banner, the size of the keel, and the numbers of leaflets are important in keys, but because the ranges of possibilities for different species frequently overlap, it is difficult to be definitive.

In general, *Astragalus* maintains a low stature, often spreading on the ground, and it has odd-pinnate compound leaves. There usually are clusters of small pink to magenta (but sometimes white or red) flowers toward the ends of stalks that grow from leaf axils, and pods are usually inflated around small seeds. Pods are too short and fat to look like a normal pea pod, and they may be divided part or all the way down the middle, forming two chambers.

Astragalus coccineus (**scarlet milkvetch**) has "brilliant red" (*coccineus*) flowers about 4 cm long, looking like little firecrackers. The plant is low growing, forming a tuft of silky-hairy leaves as a backdrop for the flowers. It likes gravelly locations, including roadbeds or shoulders, at elevations of 750–2,450 m in the western and northern deserts.

Astragalus coccineus (scarlet milkvetch)

Astragalus didymocarpus (**two-seeded milkvetch**) is a slender annual with tiny stiff hairs pressed against the stem. Flowers are whitish with a purple tinge, and fused sepals are covered with hairs, some black and some white. Pods are unusually small, only 2–4 mm long, and almost spherical; they are also two-lobed and two-chambered in cross section, as the specific epithet, *didymo-* (in pairs or two-lobed) plus *carpus* (fruit) suggests. Usually there are just two seeds, as the common name makes plain. The two desert varieties of *Astragalus didymocarpus* are distinguished by having either mostly white or mostly black hairs on the calyx.

Astragalus layneae (**Layne milkvetch**) honors Dr. Mary Katherine Layne Curran Brandegee (1844–1920), an intrepid California plant collector and botanist.

Astragalus didymocarpus (two-seeded milkvetch)

Astragalus layneae (Layne milkvetch)

Astragalus layneae is a perennial with stems that are unusually erect for *Astragalus*, and covered with coarse grayish hairs. Most calyx hairs are black with a few white, and petals are white

Astragalus lentiginosus var. *fremontii* (freckled milkvetch)

with some purple, mostly at the tips. The banner is usually light purple or lavender, and 10–45 flowers are loosely clustered along erect stem tips. Fruits are only slightly inflated, and are curved around in the shape of the new moon. *Astragalus layneae* is fairly common along sandy roadsides and washes of the Mojave Desert.

Astragalus lentiginosus (freckled milkvetch) is a widespread perennial (usually) of highly variable form. Pods generally are strongly inflated and "freckled" (*lentiginosus*) with irregular purple spots. There are 19 varieties just in California, with seven of these occurring in the deserts. Many varieties are highly localized and vulnerable to extinction due to encroaching habitat disturbances. One of the more common varieties in the Mojave Desert is *Astragalus lentiginosus* var. *fremontii,* which may be either annual or perennial, either densely or sparsely hairy, and either spreading or erect. Flowers are usually some shade of purple, and pods are strongly inflated and either thinly hairy or hairless. Stems and leaflets are a gray or silvery shade of green. If the description sounds indecisive, it simply reflects the extreme variability seen in the field. The varietal name honors John C. Frémont, one of the West's most vigorous explorers and politicians.

Astragalus newberryi var. newberryi (Newberry locoweed) is a perennial that retains its old leaf bases, making the stem appear untidy. It also has long silky hairs, more white than black on the calyx. Each cluster comprises just a few (three to eight) flowers, which are pinkish purple or white, tipped with pinkish purple. Fruits are one-chambered and densely hairy. The specific epithet and varietal name honor the geologist and paleontologist Dr. J. S. Newberry (1822–1892). This species occurs in rocky areas of the northeastern desert mountains, at elevations from 1,300 to 2,350 m.

Astragalus newberryi var. *newberryi* (Newberry locoweed)

Dalea (dalea)—According to the Jepson manuals (1996 and 2002), this genus was named after the 18th-century English botanist Dale. It has just four species in California, three of which are desert residents. *Dalea* are small annuals or perennials, dotted with dark glands. They have odd-once-pinnate leaves. Flowers break a few rules: stamens are not hidden within the keel, and there sometimes are just nine rather than 10 stamens. One desert species, *Dalea searlsiae*, has only five stamens. Although *Dalea* species have two ovules, only one of them develops into a seed, an unusual condition for a bean.

Dalea mollissima (silk dalea)

Dalea mollissima (silk dalea) is a hairy and gland-dotted ground-hugging annual. *Mollissima* means that it has the "softest" hairs, even more soft than its cousin *Dalea mollis*. Both species have whitish to lavender petals that are hard to see among long, hairy calyx lobes. They are common at low elevations (under 800 m), and most recognizable for their abundant black dotlike glands among the soft hairs.

Glycyrrhiza (licorice)—The Greek words *glykyr-* (sweet) and *rhiza* (root) describe the extremely sweet sugar found in the roots of licorice. Licorice, of course, is that smelly black candy (originally a medicine) that has been admired for over 4,000 years. The genus has about 20 species, and the one responsible for commercially produced licorice, *Glycyrrhiza glabra*, grows in a few parts of California but is native to Eurasia. It is not naturalized in the desert. California's one native species of *Glycyrrhiza* grows in many disturbed areas of the state, including the desert.

Glycyrrhiza lepidota (wild licorice)

Glycyrrhiza lepidota (wild licorice) is a perennial, generally spreading in moist disturbed sites by new sprouts from the roots. It is common along roadways and is popular with ants, which often infest it. Plants are usually extremely sticky, due to dense glandular hairs, but some populations may be free of hairs. Leaves are odd-once-pinnate with nine to 19 leaflets. Flowers are about 1 cm long and cream or greenish white, best admired from the car. The Greek word *lepidota* means "covered with small scales."

Lotus (lotus)—The genus name probably comes from the Greek *lo* (to cover), but the rationale for this origin is obscure. California is home to 33 species of *Lotus*, three of them nonnative, and there are 13 desert species, all native. In addition, there are many varieties, and identification is complicated by hybridization and overlapping features. *Lotus* species are annuals, perennials, or shrubs, without spines or prickles. Leaves are usually odd-once-pinnate, with three to many leaflets. Numbers of leaflets and the types of stipules are important in determining species. Flowers are almost always bright yellow, sometimes with reddish markings or turning red with age. One desert species, *Lotus oblongifolius,* has light yellow flowers. There are nine fused stamens and one that is separate. Fruits look like small beans, and may be dehiscent (splitting open on their own), like good beans, or they may be indehiscent. This is an important feature for identification, requiring mature fruit.

Lotus rigidus (**desert rock-pea**) is a perennial that seems like a shrub due to its *rigidus* (rigid), upright stems, but they are not woody, as required for a shrub. The stems are dark green, fairly straight, and branched in clusters. Leaves are sparse, often with just three leaflets in an almost palmate arrangement. There may be glandlike stipules or none. Flowers, up to 2 cm long, come in small groups of up to three. Beans are very straight and also *rigidus,* often with many seeds. *Lotus rigidus* is fairly common in desert flats and washes under about 1,550 m.

Lotus rigidus (**desert rock-pea**)

Lupinus (**lupine**)—This genus is named for the "wolf" (*lupus*), theoretically because these plants rob the soil of nutrients—an odd mistake, given that Fabaceae species, with their bacterial associates, typically add nitrogen to the soil. Maybe its "wolf"-like nature had more to do with killing sheep, because several species are highly toxic to livestock, especially sheep. In any case, there are about 200 species worldwide, with a concentration in western North America, where California alone has about a third of them (71), all natives. Fourteen of California's species occur in the desert.

Lupines are easy to recognize by their palmate arrangement of leaflets, generally five or more, spreading from a central point of attachment. Leaves have petioles that are often long, and large stipules are fused to the base of petioles, one on each side. The fruits are often hairy, but they still look like small beans. As they dry, they split open along both sutures, and the two sides often twist into mirror-image spirals. Desert lupine flowers grow in conspicuous spikes and are almost always colored somewhere on the spectrum of blue, purple, violet, or rose, usually toward the blue end.

Rarely an all-white specimen occurs, more common in some species than others. Often the banner has a spot of another color such as yellow, pink, or white. The spot usually turns to a darker color following fertilization of the flower, signaling bees that there is no longer any nectar available (or any reason to bring pollen). Ten stamens with fused filaments are the rule. Half are short with long anthers, and half are long with short anthers. Lupines may be annuals, perennials, or shrubs. After that, identification depends, among other things, on banner shape and coloring, precise location of hairiness on the keel, type of hairiness on leaves, and position of leaves along the stem or at the base of the plant.

Lupinus arizonicus (**Arizona lupine**) is a large annual, up to a half-meter tall. It has magenta to violet petals, the banner and

Lupinus arizonicus (**Arizona lupine**)

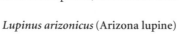

wings being more pinkish than the keel. The banner has a yellowish spot, turning dark magenta. Stems and undersides of leaves are hairy, and there are 6–10 leaflets. Arizona lupine grows in open sandy areas of the Sonoran and eastern Mojave Deserts at elevations below 1,100 m.

Lupinus concinnus (**bajada lupine, elegant lupine**) is a small annual, no more than 30 cm tall. It generally has a "neat" or "elegant" (*concinnus*) appearance. The five to nine leaflets are hairy and often almost linear. Petals are pink to purple with yellowish or white banner spots. In addition to the usual

Lupinus concinnus (bajada lupine)

spike of flowers, a few flowers appear singly in axils of leaves. The fruit is 1–1.5 cm long and hairy, with three to five seeds. Bajada lupine is common in open and burned or disturbed areas (not necessarily bajadas, a term for alluvial fans) under 1,700 m.

Lupinus excubitus (**grape soda lupine, Inyo lupine**) is a shrub or subshrub whose flowers smell distinctly like grape soda—which is really all you need to remember.

Excubitor means "watchman" or "sentinel." Variety *excubitus* grows under 2,500 m in the Mojave Desert, and the rare variety *medius* (Mountain Springs bush lupine) grows in limited portions of the Sonoran Desert.

Lupinus magnificus (**Panamint Mountains lupine**) is one of the most "magnificent" (*magnificus*) species of lupine, tall (up to about 1 m) and

Lupinus excubitus var. *excubitus* (grape soda lupine)

Lupinus magnificus (Panamint Mountains lupine), Panamint Mountains above Mahogany Flat

stately, with fragrant bluish to rose flowers and silvery leaves. Anyone tempted to collect it will be deterred by sharp stiff hairs that coat everything, including fruits, and come off easily into your skin and clothing. Look for it at elevations of 1,500–2,500 m. Although rare, growing only in northern desert mountains, it is abundant near the charcoal kilns on the road to Telescope Peak in Death Valley National Park.

Lupinus microcarpus (**chick lupine**) is an annual to 80 cm tall, but usually 20–30 cm, with the larger stems hollow. It generally has nine very narrow leaflets that are hairy only on the bottom.

Lupinus microcarpus (chick lupine)

The flower color is variable between pink, lavender, and purple. The hairy pods are less than 1.5 cm long and almost as wide, with just two seeds. Hence the specific epithet *micro-* (small) *carpus* (fruit). Maybe the fruits look like fuzzy, round baby chicks, accounting for the common name. Two varieties grow in the Mojave Desert.

Lupinus odoratus (**Mojave lupine, royal desert lupine**) has a "fragrance" (*odoratus*) like violets. Furthermore, it is stately, with royal purple flowers, and grows in the Mojave Desert. It is an annual to 30 cm tall, with seven to nine bright green leaflets that are only hairy when young. Intensely colored petals have a white banner spot and take center stage in sandy open areas under 1,600 m.

Marina—This is a small genus, named after the Aztec interpreter for Hernán Cortés, who conquered Mexico in 1521 and then ruled as its first Spanish governor. Most of the 38 species are native to Mexico, but two extend into California, one of them a desert species.

Marina species have odd-once-pinnate leaves with the even ones neatly arranged opposite one another. The flowers bear 10 stamens with fused filaments. Fruits are indehiscent (not opening on their own) and contain only one seed.

Marina parryi (**Parry dalea**) is a fragile-looking perennial with long thin stems and small sparse leaves. It may grow to 80 cm, but it is not very dense. The stems are covered with tiny stiff hairs, and usually with glands. Each leaf has 11–23 small (0.5–6 mm) leaflets that are also dotted with glands. The flowers too are

Lupinus odoratus (Mojave lupine)

small (5–7 mm), clustered together in spikes, and colored a deep purple-blue. Charles Parry (1823–1890) was a botanist who collected extensively in North America, especially in the Southwest. *Dalea*

Marina parryi (Parry Dalea)

is the genus from which *Marina* was segregated, hence the common name. *Marina parryi* grows at the lower elevations (under 800 m) of the deserts, in open washes and along roadsides.

Olneya This genus with only one species is named after Stephen Thayer Olney (1812–1878), best known for his work on sedges (*Carex*), in Cyperaceae.

Olneya tesota (**ironwood**) is a small tree with exceedingly hard, "iron" wood. *Tesota*

Olneya tesota (ironwood)

PHOTOGRAPH BY LORI PAUL

Psorothamnus arborescens (Mojave indigo bush)

Psorothamnus emoryi (Emory indigo bush)

comes from the Spanish *tieso*, meaning "stiff" or "firm," describing the inflexible nature of the hard branches. To confuse matters, another ironwood (also known as Catalina ironwood) can be found in the Rose Family (Rosaceae), in the single-species genus *Lyonothamnus*. Desert ironwood, however, is unmistakably in Fabaceae. Pods look like lumpy beans containing just a few seeds. When roasted, the seeds are said to taste like peanut butter, and were a favorite part of Indian diets. Flowers, much attended by bees, are about 1 cm long with purple wings and lighter-colored banner and keel, usually cream to pink or lavender. Ironwood has even-once-pinnate leaves with 8–19 thick leaflets arranged not quite opposite one another. The main leaf axis extends beyond the farthest leaflets, making a small point, and the leaves frequently grow in clusters. There are spines for stipules. Ironwood is often abundant in open sandy washes of the Sonoran Desert below about 600 m.

Psorothamnus (**indigo bush**)—This genus, with just nine species from southwestern North America, includes subshrubs, shrubs, and trees. *Psorothamnus* species are intricately branched and often well armed. The abundant flowers are blue to purple, sometimes with white or orange included somewhere, and embrace 10 stamens with partially fused filaments. Although there generally are two ovules per flower, only one develops into a seed. Pods are short and dotted with colorful glands. *Psoro-*, meaning "scabies" or just "scabs," combined with *thamnus*, meaning "shrub," pretty well describes this genus with its scabby-looking glands. Another distinctive feature of the genus is its strong green, legume odor, which will stay on your hands and clothes after you have handled any part of the plant. California boasts six species of *Psorothamnus*, all of them desert natives.

Psorothamnus arborescens (**Mojave indigo bush**) has three varieties, all occupying portions of the Mojave Desert. Indigo is definitely the best description of the intense deep purple-blue of the flowers. The varieties are all shrubs, *arborescent*, or "becoming treelike," in their woody branching, but still only 1 m or less in height. Leaves are odd-once-pinnate with five to seven leaflets, small features of which help to distinguish varieties.

Psorothamnus emoryi (**Emory indigo bush**) is a subshrub, generally well under 1 m and usually wider than tall. It is light gray-green due to a powdery surface on the twigs and leaves, which sometimes

Psorothamnus fremontii (Fremont indigo bush), eastern Mojave Desert, Shadow Valley

disappears with age. Glands are small—considerably less than 0.5 mm on the twigs. The odd-once-pinnate leaves are small also, with five to nine neat leaflets, the end one being largest. Flowers, deep purple dotted with white, are only 4–6 mm long, but they grow in dense spikes or almost round clusters. *Psorothamnus emoryi* will give your hands a lovely saffron-color stain in addition to the odor common to the genus. It grows in low sandy flats and washes of the Sonoran and southern Mojave Deserts. The specific epithet honors Major W. H. Emory (1812–1887), who directed the Mexican Boundary Survey.

Psorothamnus fremontii (**Fremont indigo bush**) is named to honor the explorer John C. Frémont, whose name is attached to several species and one genus of plants. His *Psorothamnus* is a silvery-looking shrub with flowers almost 1 cm long. Leaves are odd-once-pinnate. Fremont indigo bush frequents eastern desert mountains and canyons at 250–1,350 m.

Psorothamnus spinosus (**smoke tree**) is a small tree growing along low (under 400 m) desert washes. It is so light a gray-green that it appears to be just a puff of smoke. *Spinosus* describes the spines (technically thorns) that form as the tips

Psorothamnus spinosus (smoke tree)

of numerous twigs harden with age. Leaves are simple and short-lived, usually falling before flowers open. The flowers are deep purple-blue and only 6–8 mm long, but they bloom abundantly, turning the gray-green smoke to a purple haze in summer.

Examples and Uses

Fabaceae is second only to Poaceae (grasses such as wheat and rice) in its importance to the human diet. Legumes are better sources of protein than are grasses, and also are unusually high in iron and vitamins A, B, and C. The nutritional value depends in large measure on species, cooking method, the part of the plant that is eaten, and its maturity. We eat legumes as fresh seeds (green peas), as dried seeds (lentils), as sprouts (mung beans), as whole fruits (green beans), and as soy sauce, cooking oil, milk substitute, and bean curd (soy beans), to name just a few possibilities.

Nearly every part of the world where cultivation originated has long depended on legumes as important crop plants. Lentils (*Lens culinaris*) were domesticated in the Near East possibly as much as 9,000 years ago. Central and South America, separately, each domesticated lima beans (*Phaseolus lunatus*) and common beans (*Phaseolus vulgaris*) as much as 8,000 years ago. The Mediterranean region had the fava (or faba) bean (*Vicia faba*), India had the mung bean (*Phaseolus aureus*), and Asia had the soybean (*Glycine max*). As soon as trade routes were opened, crop plants traveled rapidly, often becoming staples on other continents. For example, the peanut (*Arachis hypogaea*) soon moved from its Amazonian origins to become important in Africa.

For us, garden-variety beans (*Phaseolus vulgaris*) and peas (*Pisum sativum*) are probably the most familiar examples of foods in Fabaceae. Economically, however, soybeans are far more important, and are cultivated on a large scale in many parts of the world. Soybean products are used mainly as animal fodder, but they have applications in industry as well. The Model-T Ford had bumpers made from soybeans, and soy is the basis for many modern printing inks. Clover (*Trifolium*), sweet clover (*Melilotus*), and alfalfa (*Medicago sativa*) are common forage crops for domestic grazing animals.

Do not get the idea that all Fabaceae are edible. Many are highly toxic and have never served as food sources. The rat poison rotenone, for example, is contained in the genera *Derris* and *Lonchocarpus*. Even the edible lima bean contains cyanide precursors and should not be eaten raw. Boiling in an uncovered pot releases hydrogen cyanide gas, but not in quantities large enough to be dangerous in your household air. Many other legumes, common in our diets, need to be cooked to improve digestibility or to restore the water content after seeds have been dried.

One of the most important uses of Fabaceae comes from the ability of many species to harbor nitrogen-fixing bacteria in their roots. The bacteria make nitrogen available to their hosts—acting like little internal fertilizer factories. Some Fabaceae crops produce enough nitrogen that another crop, planted subsequently, will benefit from the increased nutrient levels in the soil. For example, clovers, alfalfa, and other "cover" crops, as they are termed, may be grown during the winter in mild climates and then ploughed under to improve the soil for spring planting of grains or other nonleguminous crops.

Certain Fabaceae members have other important uses. *Acacia*s provide certain gums, as well as being drought-tolerant ornamentals. *Wisteria* and sweet peas (*Lathyrus*) also are well-known ornamentals. Rosewood used in fine furniture has nothing to do with roses, but comes from the Fabaceae genus *Dalbergia*.

FOUQUIERIACEAE

OCOTILLO FAMILY

Fouquieria splendens ssp. *splendens* (ocotillo), Anza Borrogo Desert State Park, Palm Canyon

California's spectacular member of Fouquieriaceae marks the extent of our Sonoran Desert and grows nowhere else. When present, ocotillo's tall spiny stems can hardly be missed, and in spring, clusters of bright red flowers attract tourists and hummingbirds alike.

A 19th-century French professor of medicine, Pierre-Edouard Fouquier, provides the basis for this family name and that of the lone genus (*Fouquieria*) in the family. Ocotillo is an Aztec word for "pine," with a Spanish ending that means "little."

Fouquieriaceae Icon Features

- Tall spreading branches from a very short trunk
- Neat rows of spines on older branches
- Tubular red flowers lining the tips of branches

Distinctive Features of Fouquieriaceae

Everything about this family is distinctive. Fouquieriaceae species all are spiny shrubs or trees with a profusion of bright red or yellow flowers in spikes at the ends of branches. Usually, branches are numerous and rise or spread to several meters in height from a nearly invisible trunk at ground level. One species, the boojum tree (*Fouquieria columnaris*), which has sometimes been considered a second genus (*Idria*), looks more like a tree, with its single, columnar, spiny trunk and shorter branches spreading well above ground level. California's only species, *Fouquieria splendens*, is of the invisible-trunk type with red flowers.

In this family petals are fused together into a cylindrical tube with five spreading or reflexed lobes and 10 or 20 stamens extending beyond the corolla. Stamens are attached at the base to the corolla. There is just one pistil with a superior ovary.

Leaves of Fouquieriaceae members are small (generally 1–3 cm), more or less oval, hairless, a bit fleshy, and extremely responsive to changing moisture conditions. They come in two different

Fouquieria splendens ssp. *splendens* (ocotillo)

configurations. Primary leaves sprout from bare branches and grow to full size within a few days of a good rain. These leaves are located all along each of the numerous branches, forming a regular pattern of single leaves. After the soil has dried, the central vein of the primary leaf blade and its petiole harden into a spine, while the wide part of each leaf blade dries and falls away. Thus, with the work of photosynthesis no longer feasible, the primary leaf turns into a protective device. Subsequent rains produce secondary leaves that grow in small clusters in the axil of each of the protective spines.

Similar Families

The growth form in this family is so distinctive that there is no danger of confusing it with any other California desert family, genus, or species. There is, however, a fairly common house or garden plant with similarly spiny stems, red flowers, and a leaf pattern like that of ocotillo. This native of Madagascar, called crown of thorns, is in the Spurge Family (Euphorbiaceae) and has adaptations for arid conditions similar to Fouquieriaceae but is otherwise quite different. If in doubt, pull off a leaf: the spurge has milky sap.

Family Size and Distribution

All 11 species of the single genus *Fouquieria* are native to Mexico or the southwestern United States. The only species in California is *Fouquieriaceae splendens*, and it is often a dominant feature of the Sonoran Desert. It grows only in Sonoran Desert and is widespread there, so if you see it, you know where you are.

California Desert Genus and Species

Fouquieria (**ocotillo**)—The genus carries the family names, both scientific and common.

Fouquieria splendens **ssp.** *splendens* (**ocotillo, candlewood**) stands out, unique in any season, in its desert habitat. During dry periods it is leafless and flowerless, appearing to be a huge loose clump of long spiny poles sticking out of a hole in the ground. This formation generally towers over surrounding shrubs and cacti. Ocotillo changes quickly to leafy, spiny poles following sufficient rain, but it retains the unique profile. During mid- to late spring of a good year, long (20 cm or more), dense clusters of bright red flowers appear on the ends of nearly every branch. These flower spikes could be likened to a flame atop a candle and may account for the common name candlewood. Certainly, the Latin specific epithet, *splendens*, is appropriate to the flowering appearance of this magnificent plant.

Examples and Uses

Nowadays, California's *Fouquieria splendens*, in flower, is probably most frequently used by migrating birds, especially hummingbirds moving north for the summer. However, native people have found many uses for different parts of the plant. The thin straight branches have been incorporated into shelters and are especially effective as fences because of the dense spines. Furthermore, branches pushed into the ground may root and grow additional branches, forming a virtually impenetrable barrier—a sort of desert barbed wire. Sometimes adobe bricks are reinforced with ocotillo stems, acting like re-bar. Flowers can be eaten raw, and flowers together with seeds are soaked in water to make a type of tea. A waxy substance extracted from stems has been used as leather dressing.

Phacelia campanularia ssp. *vasiformis* (desert Canterbury bells), Joshua Tree National Park, near the south entrance

HYDROPHYLLACEAE

WATERLEAF FAMILY

Some of the most spectacular desert blooms following a good rainy season are produced by annuals of Hydrophyllaceae. These lovely flowers, usually blue to lavender, generally grow along a coiled stem tip. Often there are several species growing in close proximity, as if they were in your garden. However, do not be tempted to pick them, as many have hairs and glands that will irritate your skin and disgust you with their odor.

The genus from which this family gets its name, *Hydrophyllum*, derives from the Greek: "water" (*hydro-*) "leaf" (*phyllum*). Clearly, the desert representatives of this family were not consulted when the name Hydrophyllaceae was devised. In fact, the moisture-loving genus *Hydrophyllum* is not even found in the desert.

Hydrophyllaceae Icon Features

- Tubular corolla with five spreading lobes
- Five long calyx lobes
- Five stamens usually extending beyond the corolla tube
- Style forked and usually long

Distinctive Features of Hydrophyllaceae

Coiled spikes of small flowers are characteristic of many of the most abundant desert genera of Hydrophyllaceae, but this is also true of the Borage Family (Boraginaceae). To distinguish Hydrophyllaceae, look for a forked style and a blue to lavender corolla (though here too, of course, exceptions exist). Plants are usually hairy and glandular, and they often have a bad odor, especially if touched or crushed. Many species have irregularly lobed leaves or rumpled-looking leaves, though this is not universal.

Hydrophyllaceae species may be annuals, perennials, or shrubs, and our deserts have them all. Leaves are very diverse, ranging from entire to pinnately compound, with lobed ones especially common. Furthermore, leaves may be alternate or (rarely) opposite in arrangement, and growing at the base of stems or arranged along stems, or both. Glands and hairs, although common, are not consistent, and certainly are not unique to the family.

Flower structure, at least superficially, is shared with several other families. Symmetry is radial, and parts are generally in fives, with five calyx lobes, five corolla lobes, and five stamens attached to the corolla between lobes. There may be small appendages in various places, usually requiring dissection and a hand lens to see clearly. The ovary is superior (at least in desert genera) and has a single chamber, although certain configurations may result in the appearance of two to five chambers. If all these common features are present, check the style carefully. Usually it is two-lobed or two-branched, but sometimes just the stigma is two-lobed. Occasionally there are two styles. Each fruit contains two to many seeds.

Similar Families

Boraginaceae and Hydrophyllaceae are often confused because of the similar way flowers grow along a coiled stem tip, with the oldest ones maturing at the base of the coil as it unwinds and new flowers open toward the tip. The most reliable difference between the families is that Boraginaceae fruits are deeply lobed into four separate nutlets, each with a single seed, whereas Hydrophyllaceae fruits are round to oval, usually with a single cavity filled by two to many seeds. A more obvious, if less reliable, difference between the families is corolla color, typically in the blue to pink range for Hydrophyllaceae flowers (think of blue "water" [*hydro*]) and either white or yellow for Boraginaceae flowers (generally true of desert species but not others)—though naturally, there are exceptions. Another clue is that deeply lobed or compound leaves are common to Hydrophyllaceae species but do not occur in the Boraginaceae members that inhabit the desert.

The Phlox Family (Polemoniaceae) is closely related to Hydrophyllaceae. Flowers are similar, and both families have hairy and glandular annuals. Phlox Family flowers have a style with three branches instead of two, however, and flowers do not grow in coiled spikes.

The Nightshade Family (Solanaceae) also has many flower features in common with Hydrophyllaceae, but lacks coiled spikes of flowers. Nightshade Family flowers have a single style, although the stigma may be two-lobed. In general, Nightshade Family corollas have a tube that is narrow at the base and roughly trumpet-shaped, whereas Waterleaf Family corollas generally spread near the base and are saucer- to bell-shaped.

Finally, the Morning Glory Family (Convolvulaceae) has a flower structure similar to all of the above families, with a style that may be two-lobed in its very few desert species. Morning Glory Family members do not have flowers in coiled spikes, though, and most desert species are subshrubs with rather large white flowers or are twining like vines—features rarely found in Hydrophyllaceae.

Family Size and Distribution

There are only about 20 genera and 300 species of Hydrophyllaceae worldwide, but they are heavily concentrated in the western United States and quite abundant in California's deserts. Over half the species in the family are in the genus *Phacelia*. *Nama* is another large genus. The remaining species are distributed among much smaller genera.

California has 13 of the 20 genera of Hydrophyllaceae, and nearly all species present in California, outside of our gardens, are natives. California's deserts have 10 genera, including more species of *Phacelia* than you care to know.

Most desert species of Hydrophyllaceae are annuals that survive long seasons, or even years, of drought as seeds. After good rains they germinate virtually everywhere, especially in loose sandy soils and in the shelter of shrubs such as creosote bush. Their flowers can form a sea of blue. *Phacelia* species, in particular, grow prolifically in sandy washes and spots that accumulate runoff.

Several species of Hydrophyllaceae are abundant in recently burned areas. In at least some cases, the germination of seeds requires exposure to nitrogen compounds that occur in smoke. This trigger assures that seeds will germinate in open places where competition from other plants has been reduced, making successful production of the next generation of seeds likely. In air basins where pollution mimics the triggering effect of smoke, seeds may be wasted by germinating where conditions are not suitable for successful growth and production of a new crop of seeds. This mechanism could result in loss of these species from areas with air pollution.

California Desert Genera and Species

Glandular annuals with blue flowers growing in coiled spikes are bound to be Hydrophyllaceae, very likely of the genus *Phacelia*. Other genera, such as *Nama*, have predominately pink flowers on low-spreading, heavily branched plants. White flowers, though less common, break the color codes of both *Phacelia* and *Nama*. Another exception is the yellow coloring of *Emmenanthe*.

In general, perennials of the family look very similar to their annual relatives, but can be distinguished by the presence of dried stems that were fresh the previous year. The few shrubs of the family have flowers with the usual Hydrophyllaceae features, growing in clusters at the ends of upper branches that retain their leaves throughout the year.

SIMPLIFIED KEY TO DESERT GENERA OF HYDROPHYLLACEAE

Several genera of Hydrophyllaceae found in California's deserts are identified by readily observable unique features. After that, genera must be determined by a close examination of flower styles and calyces, or by a combination of corolla color and leaf features. More obscure features, such as root type and number of ovary chambers, are frequently used in keys that cover the whole family, but for California's desert genera these complexities are not necessary.

1. Shrub with evergreen leaves and shredding bark, 1–2 m tall, leafy in all seasons (2 spp) .... *Eriodictyon*
1' Annual or perennial, rarely an ill-scented subshrub.
  2. Flowers solitary on leafless stems from ground level. Leaves all basal ..... *Hesperochiron pumilus*
  2' Flowers several to many on coiled stem tips or singly in leaf axils, generally several per stem. Some leaves along stems.
    3. Calyx lobes unequal with 3 greatly enlarged into heart shapes .... *Tricardia watsonii*
    3' Calyx lobes almost equal, or at least not heart-shaped.
      4. Styles 2, separate from each other from the ovary.
        • Tall perennial or subshrub, densely glandular and ill-scented. Leaf blade 4–30 cm long .... *Turricula parryi*
        • Low annual, hairy, some glandular, no distinctive odor. Leaf small (6 spp) .... *Nama*
      4' Style 1, but often with 2 long lobes.
        5. Calyx generally with reflexed appendages between lobes; if appendages absent, corolla white with a purplish or greenish dot near the center of each lobe. Leaves and stems often with strong bristles but not hairy or glandular. Stems generally brittle and often resting on the ground or on other plants.
        • Fruit spherical with bristles. Stems many-branched, spreading. Calyx with (rare) or without (common) reflexed appendages. Corolla blue to purple with dark marks (Whipple Mountains only) or white with a purplish or greenish spot near the center of each lobe (2 spp) .... *Pholistoma*
        • Fruit ovoid with hairs but not bristles. Stems generally upright. Calyx with reflexed appendages. Corolla blue or blue-veined with black or white center ... *Nemophila menziesii*
        5' Calyx without reflexed appendages; corolla not white, with a purple dot near the center of

each lobe. Leaves and stems frequently glandular and hairy, sometimes with thin bristles. Stems generally upright, sometimes growing through shrubs.

- Corolla pale yellow, pendulous, and bell-shaped .... *Emmenanthe penduliflora*
- Corolla white, or bluish or creamy white, and small (less than 6 mm). Leaves finely dissected like ferns and pleasantly scented (2 spp) .... *Eucrypta*
- Corolla blue, lavender, violet, or purple, rarely white, sometimes with yellow tube or white markings. Leaves dissected or not, unpleasantly or not scented (34 spp) .... *Phacelia*

Emmenanthe—This single-species genus has a Greek name meaning "abiding" or "faithful" (*emmen-*) "flower" (*anthe*) because the corolla remains attached as it withers and the fruit matures. Most other members of Hydrophyllaceae loose their corollas early, just after flowers have been pollinated or as new flowers open on the stem above.

Emmenanthe penduliflora (**whispering bells**) has a species epithet describing its "hanging" or "pendulous" (*penduli-*) "flowers" (*flora*). They are also bell-shaped, and possibly even whispering. Most notably, the corolla is pale yellow, which is unusual for the family. Whispering bells is an odorous and sticky glandular annual, usually with soft, short hairs. Leaves are long and narrow with lots of regularly sized teeth or lobes along each side. Whispering bells grows in rocky or sandy soils under 2,200 m. Though seldom found in abundance, it is widespread in both the Mojave and Sonoran Deserts.

Emmenanthe penduliflora (whispering bells)

Eucrypta—This genus has just two species, both of which can be found in California's deserts, although they are not easy to spot. The Greek name, meaning "well" (*eu-*) "hidden" (*-crypta*), refers to small seeds in a bristly fruit, but the flowers also tend to be hidden by both their small size and their habitat, which includes rocky crevices.

Eucrypta chrysanthemifolia (**chrysanthemum-leaved eucrypta**) has a species epithet referring to leaves that resemble those of *Chrysanthemum*, a genus in the Sunflower Family (Asteraceae). Both *Eucrypta* species have a corolla only 2–4 mm long and either white or bluish white with some yellow inside the tube. Both also have weak stems, often resting on rocks in crevices. The lower leaves are opposite and much larger than the upper alternately arranged leaves. In *Eucrypta chrysanthemifolia*, larger leaves are deeply lobed, with each lobe again lobed, giving a somewhat fernlike appearance.

Eucrypta chrysanthemifolia (chrysanthemum-leaved eucrypta)

Foliage is scented, like many others in the family, but not unpleasantly so. *Eucrypta micrantha* has deeply lobed leaves but they are not fernlike and they are sparse along upper stems. Both species grow in protected rocky locations at elevations under 2,500 m.

Nama (purple mat)—This genus has about 55 species. Its Greek name means "spring" or "stream," but that is not where you will find desert species, which generally prefer dry sandy flats. Plants, though low growing, are more mounded than matted and generally more pink than purple, but still, the common name can be a memory jogger. Most species are hairy annuals with lots of thin branches and simple, small leaves that tend to be concentrated near the tips of branches next to flowers. Many *Nama* species have a trumpet-shaped corolla, and stamens are usually of irregular length and attached at various levels within the corolla tube.

Nama demissum (purple mat) is common in sandy or gravelly flats under 1,600 m. Its numerous forked branches form a low mound that may be completely covered with bright rose-pink flowers. The spreading part of each corolla is almost 1 cm across, making for a stunning overall effect when

Nama demissum (purple mat)

Phacelia affinis (purple-bell phacelia)

Phacelia campanularia ssp. *vasiformis*
(desert Canterbury bells)

Phacelia crenulata (notch-leaved phacelia)

the plant is large. In dry conditions, plants may be tiny and have just one or two smaller flowers. The Latin *demissum* means "low" or "humble," an apt description of the plants' posture but not of the bold color of the flowers.

Phacelia—This genus name means "bundle" or "cluster" in Greek, referring to the tight formation of flowers in coiled spikes. In some species, coils are quite loose with only a few flowers, but it is more common for them to be densely packed with numerous flowers. There are 93 species of *Phacelia* in California, 34 of which are represented in our deserts. Most desert species are annuals with glandular hairs and bluish to purple flowers, but there are plenty of exceptions. The few desert *Phacelia* species with white flowers may be confused with *Cryptantha*, in Boraginaceae. *Phacelia* corollas, spreading from the base, are saucer or bell-shaped, and styles are distinctly two-lobed. Stamens are generally equal in length and originate from like positions on the corolla. Often style lobes and anthers extend well past the corolla, helping in identifying to species. Other important features in species keys include the amount of division of leaves; numbers of seeds per fruit; size, color, and shape of the corolla; length of calyx lobes in fruit; and position and abundance of glands.

Phacelia affinis (**purple-bell phacelia**) could just as easily be named white-bell phacelia due to the range of corolla colors. Its specific epithet, *affinis*, means it is "related to" or "close to" other *Phacelias*—of which there are many. *Phacelia affinis* is a small annual, often under 10 cm but sometimes growing to 30 cm, mainly in heavily sandy soils under 3,400 m in both deserts. It has oblong leaves 1–7 cm long with deep, fairly regular lobes.

Phacelia campanularia (**desert Canterbury bells**), with its large (15–40 mm) bright purplish-blue corollas, is probably the most attractive *Phacelia* and enjoys a good reputation in wildflower gardens as well as in the desert. The distinct bell-like shape of the corolla is noted in the specific epithet, which means "resembling *Campanula* [bellflower]". The subspecies *campanularia* (predominant in the western Sonoran Desert) retains the classic bell-shaped corolla, while the subspecies *vasiformis* (of the Mojave and northern Sonoran Deserts) has a V-shaped corolla tube, like a vase. Both subspecies like open sandy areas under 1,600 m.

Phacelia crenulata (**notch-leaved phacelia**) has lovely deep violet-purple flowers with a bit of white in the

Phacelia distans (wild heliotrope) in Red Rock Canyon State Park

Phacelia pedicellata (specter phacelia)

throat. Stems are thick, generally reddish, little branched, and usually densely glandular. Leaves are deeply divided or "scalloped" (*crenulata*) into lobes or leaflets that are smallest away from the leaf tip, which remains intact. You will not be tempted to pick these after you discover the odor—like skunk with syrup. *Phacelia crenulata* commonly grows on sandy and gravelly slopes and washes under 2,200 m.

Phacelia distans (**wild heliotrope**) is one of the most common *Phacelia* species throughout our deserts. In wet years, its light blue to lavender flowers are abundant in clay, sand, gravel, and rocky soils under 2,100 m. Even in dry years, look for *Phacelia distans* growing up through shrubs and flowering in the protection of a thin creosote bush canopy. Its leaves are compound, with each leaflet toothed, lobed, or even bearing its own toothed leaflets, making them look like hairy ferns. *Distans* means to "stand apart" or to be "distant." The common name should not invite confusion with species in the Borage Family of the genus *Heliotropium*.

Phacelia pedicellata (**specter phacelia**) has beautiful, neatly arranged pink to blue flowers and a foul odor. *Pedicellata* refers to the short (1–2 mm) pedicels attaching each flower to the main stem. Stems are stout, with few branches and large rumpled round leaves. Lower leaves may be compound with three to seven rounded leaflets. The whole plant is glandular and hairy. It likes gravel or rocks of washes and canyons below 1,400 m.

Phacelia perityloides (pearl-o'rock)

Phacelia perityloides (pearl-o'rock) is the pearl of the rocks—one of very few desert *Phacelia* with a white corolla. It grows only in rock crevices of northern desert canyons. Its leaves "resemble" (*-oides*) *Perityle*, an unrelated genus in the Sunflower Family (Asteraceae) with similar leaves. The leaf blade is almost round, slightly toothed, and shorter than its petiole.

Pholistoma (fiesta-flower)—The Greek name for this genus means "scaly" (*pholis-*) "mouth" (*stoma*), referring to scalelike appendages of the corolla tube. All three species of the genus grow in California, with two found in our deserts. *Pholistoma* species are slightly fleshy annuals that branch profusely and spread over shrubs and rocks, usually in shaded locations. Leaves are deeply lobed, with thin strips of blade on both sides of major veins.

Pholistoma membranaceum (white fiesta-flower) has white corollas, less than 1 cm across. There is usually a purple spot near the center of each corolla lobe, and lobes may be hairy. Stems are not hairy but usually have bristles or hooked prickles that you might not notice until you have had a

Pholistoma membranaceum (white fiesta-flower)

direct encounter. Even the leaves have bristles, as do the round fruits. The specific epithet means "pertaining to" (*-aceum*) a "membrane" (*membran-*), a reference to the membranous tissue in the U-shaped areas between the lobes of the leaves, like the thin section of skin between spread fingers.

Tricardia—This is a single-species genus with a Greek name that describes the "three" (*tri-*) enlarged calyx lobes that look like "hearts" (*cardia*).

Tricardia watsonii (three hearts)

Tricardia watsonii (**three hearts**) is named for Sereno Watson (1826–1892), who assisted and then succeeded Asa Gray directing the Harvard University herbarium bearing Gray's name. Watson was principal author on two volumes about the botany of California. This odd plant named for Watson is noticeable because of the three papery reddish-green sepal lobes that expand in the shape of hearts—or better, playing-card spades, their stems held tightly together—below each bell-shaped, cream-colored corolla. Calyx lobes persist and enlarge after the corolla falls off and the fruit matures. The other two calyx lobes are still there but are tiny in comparison to their heart-shaped neighbors. *Tricardia watsonii* is found on sandy and gravelly slopes, usually around shrubs, at elevations of 100–2,300 m.

Examples and Uses

Several species of *Phacelia*, *Nemophila* (baby blue-eyes), and *Pholistoma* are common garden annuals. *Phacelia* species in particular are good bee plants, and in Europe *Phacelia tanacetifolia* is cultivated for honey-producing bees. *Eriodictyon* leaves sometimes are boiled into tea for use as a cough remedy. Some species of *Hydrophyllum* are reported to have edible leaves.

Krameria erecta (pima rhatany)

KRAMERIACEAE

RHATANY FAMILY

K rameriaceae members seem to be the orchids of America's dry deserts, but only because their purple to magenta flowers resemble small orchids. In fact, these parasitic plants are more closely related to beans than orchids.

Johann Georg Heinrich Kramer (1684–1744), an Austrian army physician and botanist, is the likely inspiration for the genus name *Krameria*, which has been in use since it was coined by Linnaeus in 1758. *Krameria* species were recently moved into a family of their own, necessitating the family name of Krameriaceae. Rhatany is the common name applied to the genus by Europeans arriving in the New World.

Krameriaceae Icon Features

- Five magenta sepals dominating in bilateral symmetry
- Three small upright petals above the center
- Four stamens with anther pores for pollen release

Distinctive Features of Krameraceae

Krameriaceae members may be perennials or shrubs, but the two species in California are both densely branched shrubs, less than 1 m in height, with a dark purplish cast mixed with grayish green. In both cases, flowers have bilateral symmetry, are about 1 cm across, and are deep rose or magenta to purple. They look almost like tiny orchids covering this coarse, woody shrub.

Although rhatanies have only a distant relationship to orchids, they are similar in that they develop sepals as a significant part of the colorful flower structure. In orchids, sepals are usually the uppermost and side appendages of the flower and are generally larger than petals and colorful. The four or five sepals (five in our deserts) of Krameriaceae species appear, at first, to be petals, since they, too, are larger than the petals and more brightly colored. The five sepals are arranged somewhat like the petals of several genera, such as *Senna* and *Cercidium*, in the Pea and Bean Family (Fabaceae), which probably accounts for the early assignment of *Krameria* to that family.

The real petals of *Krameria* consist of three small upright flags just above the center of the flower, plus two glands next to the ovary. The glands secrete oil that attracts certain bees. There are four stamens in *Krameria*, and the anthers, rather than splitting open lengthwise as occurs in most flowers, develop a small hole at the end. Bees that come for oil shake mature pollen out of the anthers and carry it to the next flower. This pollen-releasing mechanism also is shared with *Senna* in the Pea and Bean Family.

Petals

Sepals

Petals much smaller than sepals.
Pima rhatany (*Krameria erecta*)

Rhatany Family (Krameriaceae) sepals resemble tiny orchids.

The superior ovary of *Krameria* has two ovules, but only one develops into a seed. The fruit is not a legume (bean), which is what eventually ruled *Krameria* out of the Pea and Bean Family. Only a bean can be a bean. *Krameria* fruits do not even look like beans; they are quite round and covered with spines for an overall diameter less than 1 cm.

Krameria species harbor a secret in their root systems: they are parasites on the roots of other plants, sucking nutrients from their neighbors. They also have a backup system of photosynthesis, using chlorophyll (the usual green photosynthetic pigment) in their own leaves. Therefore, the plants do not lack green coloration, usually the most obvious sign of parasitism, nor are they entirely dependent on their parasitic capabilities. *Krameria* leaves are small and alternately arranged, without either petioles or stipules; they are usually hairy and sometimes glandular. None of these features reveals the parasitic nature of the genus.

Similar Families

The family most closely related to Krameriaceae is Polygalaceae (Milkwort Family), but there are several obvious differences between their desert representatives, and there are only two species per family to worry about. Both of the desert milkworts are of smaller stature than *Krameria*—even the woody one does not exceed 20 cm. Milkwort flowers, moreover, have cream or yellow sepal or petal members, whereas *Krameria* are almost entirely magenta, and milkworts have six or eight stamens rather than just four. The overall appearance is quite different, too, so even if you are in the limited areas of desert milkwort distribution, confusion is unlikely.

Although several similarities between *Krameria* and members of the Pea and Bean Family have been noted, the differences are far greater. Most obviously, flowers of *Krameria* are magenta to deep rose, whereas flowers of the Pea and Bean Family that resemble *Krameria* in shape (the subgroup Caesalpinioideae) are yellow. Also, most Pea and Bean Family members have compound leaves, made up of several to many leaflets, whereas the leaves of Krameriaceae species are simple.

Family Size and Distribution

Krameriaceae has only the single genus, *Krameria*, which is found only in the Americas, particularly in tropical zones, although several species are widespread in warm arid regions. There are 17 species, only two of which occur in California. The California natives are limited almost exclusively to deserts of California, Nevada, Texas, and New Mexico.

California Desert Genus and Species

The two species of *Krameria* in California look very similar and have a similar distribution. When in bloom, they are easy to spot because of the magenta flowers accentuated by the dark purplish-green background of the densely branched shrubs they cover. Shrubs are usually about 50 cm high and are often spread liberally across dry rocky slopes.

Krameria (**rhatany**)—The same names apply to the family and its only genus. Stems, leaves, flowers, and fruits of the two California desert species are quite similar at first glance, but there are distinctive differences.

Krameria erecta (**pima rhatany, purple heather**) has very narrow, almost linear leaves and blunt twigs that are pointed upward or "standing upright" (*erecta*). In coloring, *Krameria erecta* is similar to purple heather in the Heath Family (Ericaceae), though otherwise the two plants bear no resemblance to each other. Sepals are wide, with the upper four overlapping their edges and the bottom one cupped forward like a landing platform for bees. The three flag petals are also broad and slightly fringed across the top. The bases are fused together. The fruit, which is not quite spherical, has spines with tiny barbs scattered along their lengths, or occasionally there are no barbs.

Krameria grayi (**white rhatany**) is called white because of the fine, reflective hairy quality of leaves and young stems. The specific epithet honors Asa Gray (1810–1888), an important American botanist who started Harvard University's herbarium. On white rhatany, the branches are spreading and spine-tipped. Flower sepals are slim and often curved backward so that the flower looks streamlined and headed into the wind. Fruits are spherical, and their spines have barbs just at the tips, like an anchor on a long arm. White rhatany favors limestone soils but is widespread throughout the central and southern Mojave and Sonoran Deserts.

Examples and Uses

Most uses of *Krameria* are unusual, but localized and of limited economic significance. Rhatany roots have been the source for dyes, tanning agents, and toothpaste additives, among other things. The astringent properties of *Krameria triandra* roots have been valued in mouthwashes and used to reduce tooth decay. *Krameria radix* roots have been used to treat chronic diarrhea. Several species produce dyes with yellow, brown, or reddish coloring, used mainly in Mexico.

There are no cultivated examples of this family, but it is widespread in our deserts. Just go see for yourself.

Krameria erecta (**pima rhatany**)

Krameria grayi (**white rhatany**)

Salvia dorrii (purple sage), Mojave Desert north of Kelso

LAMIACEAE (LABIATAE)

MINT FAMILY

Second only to the desert, an herb garden is a good place to look for Lamiaceae. Almost all species share the quality of being aromatic, like mint, rosemary, oregano, lavender, and sage. Odors are familiar to us in a variety of products from spaghetti sauce to household cleaners. Of the famous quartet "parsley, sage, rosemary, and thyme," only one is not a member of the Mint Family. Who is the imposter?

Cowboy movie enthusiasts as well as botanists will want to make the distinction between sagebrush, in the genus *Artemisia* of the Sunflower Family (Asteraceae), and true sage of the genus *Salvia* in the Mint Family. Although both shrubs are aromatic, true sages have much prettier flowers than sagebrush, and are better represented in the Mojave and Sonoran Deserts. In his famous novel *Riders of the Purple Sage*, Zane Gray did not clarify whether he referred to the distant purple haze of *Artemisia* or the lovely purplish flowers of a *Salvia*, but botanists need to be more precise.

Lamium is an ancient Latin name and the genus name for dead nettle, native to Europe and Eurasia. Only recently has this name been applied as the root for the family name. Older books use the family name Labiatae, referring to the characteristic *labia*, or "lips," of the corolla.

Lamiaceae Icon Features

- Stems generally square in cross section
- Leaves opposite, in pairs, with alternate pairs perpendicular to each other
- superior ovary deeply four-lobed, maturing to a fruit of four nutlets

Distinctive Features of Lamiaceae

A strong aromatic odor plus a four-sided stem with opposite leaves are usually the first indications of Lamiaceae. Plants may be annuals, perennials, or shrubs (trees do not occur in our deserts), with or without hairs of various sorts and in various places, including even the corolla. Each pair of opposite leaves is positioned perpendicular to the pair below, so every other pair is in the same alignment. This configuration results in four columns of leaves, usually with a fair amount of stem showing between leaf pairs. Leaves are simple but may be variously toothed or occasionally deeply lobed or spine-tipped, and they are frequently glandular.

Flower clusters on Lamiaceae members are often distinctively spherical and arranged at intervals along a stem in a kind of poodle cut. Each cluster of flowers has bracts beneath, usually resembling leaves. In other species, flowers originate from leaf axils, but may be so congested that they form a sphere and obscure the pair of leaves from whose axils they are growing. Because the sets of bracts or leaf pairs are widely spaced along a stem, the flower clusters are also. Sometimes there are only a few flowers in a cluster or the flowers hang loosely and the spherical look is absent. Sometimes the clusters are just at the ends of branches, a few leaflike bracts seeming to hold the cluster in place.

Most flowers in Lamiaceae are bilaterally symmetrical, although some genera have radial symmetry, sometimes just a bit uneven. A bilateral corolla, tubular at the base, usually has an upper lip that is entire or two-lobed and either flat or cupped, but sometimes there is no upper lip at all. The lower lip is generally three-lobed. Radially symmetrical corollas are tubular at the base with five lobes. Calyx lobes usually are five in radial symmetry, but bilateral symmetry is also common.

Stamens, though variable, are at least highly visible, usually extending beyond the corolla lips. Most genera have four stamens, usually in unequal pairs, but there also can be two stamens along with two **staminodes**. These are nonfertile versions of stamens, usually smaller than the virile ones and with tiny or nonexistent anthers. There may also be two stamens and no staminodes.

The ovary of Lamiaceae flowers is superior and four-lobed (usually deeply lobed), ripening into four **nutlets** at maturity, each with a single seed. A single style originates from the base between ovary lobes. Two stigmas represent the fact that there are technically just two ovary chambers, each of which is deeply divided, accounting for the four lobes.

Similar Families

The Vervain Family (Verbenaceae) is a close relative of mints and mimics some of their most obvious stem, leaf, and flower features. Only three genera with five species of vervains occur in our deserts, but careful examination may be needed to avoid confusion with mints. Two genera of Verbenaceae, *Aloysia* and *Phyla*, have only two nutlets as mature fruits, rather than the four of Lamiaceae, and their calyces are two- or four-lobed

Poodle-cut arrangement of flower clusters with bracts.
Blue sage (*Salvia dorii*)

Sage Family (Lamiaceae) flowers often grow in almost spherical clusters along stem tips.

instead of five-lobed. The third genus, *Verbena*, has four nutlets plus a square stem and opposite leaves like Lamiaceae, but the two desert species are not aromatic. Also, the style of *Verbena* species is attached to the tip of the ovary rather than to the base between lobes as is the case for most Lamiaceae members. All five of the Verbenaceae species have fairly elongated spikes of flowers rather than the more compact spherical clusters characteristic of Lamiaceae; due to the variability of flowering arrangements in Lamiaceae, however, this alone is insufficient to make the distinction.

Fruits of Borage Family (Boraginaceae) members are very similar to those of Lamiaceae members, with four nutlets and a style that usually is attached at the base of the ovary between the four lobes. Otherwise, though, the families are quite different. Borages usually have flowers in coiled spikes, an arrangement never seen on a mint. Borages are seldom aromatic, whereas mint leaves usually are. Finally, and most tellingly, borages do not have square stems and opposite leaves or bilaterally symmetrical flowers, so even if you happen to look first at the nutlets, you will never end up mistaking a borage for a Mint Family member.

At first glance, some Figwort Family (Scrophulariaceae) members having opposite leaves and bilaterally symmetrical flowers with four or two stamens look like mints. Again, however, there are several fairly reliable differences. First, with rare exceptions, figwort stems are not square; second, figwort ovaries are not deeply four-lobed maturing to four nutlets; and third, if there is a staminode, it is alone, not paired. Finally, figworts are not aromatic.

An uncommon shrub of the Buddleja Family (Buddlejaceae) known as Panamint butterfly bush (*Buddleja utahensis*) could confuse you at first look, with its opposite pairs (or whorls) of leaves and spherical clusters of flowers in leaf axils. Examination of the flowers will reveal that they are unisexual and radially symmetrical with four lobes rather than the five of Lamiaceae. Fruits also are different and contain many seeds.

Family Size and Distribution

Lamiaceae is a large worldwide family of about 200 genera and 5,500 species. Many members have been cultivated as herbs or ornamentals for centuries and are broadly distributed by man. *Salvia* (sage) is the largest genus, with about 900 species, and is well represented in California and our deserts. *Thymus*, which includes the culinary herb thyme, has over 300 species, none in California. Other large genera, with about 300 species each but with small California representation, are *Hyptis* (desert lavender), *Scutellaria* (skullcap), and *Stachys* (hedge nettle).

California has 27 genera of Lamiaceae, seven of which are entirely nonnative. Some of these are escaped herbs, like oregano (*Origanum vulgare*) and horehound (*Marrubium vulgare*). Twelve of California's 27 genera are found in the deserts, including *Marrubium* with its single nonnative species, which occasionally survives in desert mountains. The genus *Mentha*, largely nonnative to California, contributes one nonnative species, spearmint (*Mentha spicata*), as an uncommon inhabitant of our desert mountains. Except for these two exotics, all other California desert Lamiaceae species are native.

No good generalization can be made as to which parts of the desert Lamiaceae members like best. There are species to fit almost all habitat types.

California Desert Genera and Species

Opposite leaves, often combined with spherical clusters of flowers, can help you spot potential Lamiaceae species, even at a distance. Some may be fairly tall shrubs, while others are low-growing annuals, but the pattern of opposite leaves and branches is similar. Flower color for desert species is almost always shades of blue to purple or white.

SIMPLIFIED KEY TO DESERT GENERA OF LAMIACEAE

Identification of genera in Lamiaceae requires a close look at the flowers, which often have very pretty and interesting bilaterally symmetrical corollas—easily distinguished from the ones with radial or almost radial symmetry. Symmetry of the calyx is also used in distinguishing among genera. One of the most important features is the number of stamens plus staminodes, which can be 4+0, 2+2, or 2+0. The genus *Salvia* has so much internal variability in features that it may be necessary to examine the attachment of anthers to the filament to be certain of genus.

1. Corolla with almost radial symmetry, 5 lobes of almost equal size.
  2. Flowers with 2 fertile stamens, annual .... *Monarda pectinata*
  2' Flowers with 4 fertile stamens, annual or perennial.
     • Leaf odor of spearmint, as in chewing gum. Flowers in leaf axils .... *Mentha spicata*
     • Leaf odor aromatic but not mintlike. Flowers just at branch tips in compact bracted clusters
     (3 spp) .... *Monardella*
1' Corolla with bilateral symmetry having an upper lip (usually) and a lower lip, each with or
  without lobes.

3. Shrubs.

  4. Flowers with 4 fertile stamens; 1 pair may be smaller. Opposite branches usually broadly spreading.

    • Calyx maturing to inflated translucent mini–paper bag. Shrub rounded, densely branched with short, rigid twigs like spines .... *Salazaria mexicana*

    • Calyx gray-green with 5 nearly equal lobes. Shrub tall and lanky, young branches thin, flexible, and covered with hairs branched in a star pattern .... *Hyptis emoryi*

  4' Flowers with 2 fertile stamens, with or without tiny staminodes. Branches spreading or not.

    • Calyx almost radial. Flowers 2–6 per cluster, in leaf axils .... *Poliomintha incana*

    • Calyx 2-lipped, each lip generally with lobes. Flowers usually many per cluster in tight spheres. Stamens with 1 or 2 anther sacs; if 2, 1 is bigger, and both hang by a thin thread (7 spp) .... *Salvia*

3' Annuals to perennials.

  5. Flowers with 2 fertile stamens.

    • Corolla with upper lip larger than the lower (2 spp) .... *Hedeoma*

    • Corolla with lower lip larger than the upper (3 spp) .... *Salvia*

  5' Flowers with 4 fertile stamens.

    6. Calyx 2-lipped, lips not lobed. Corolla white with lower lip mottled blue ....
    .... *Scutellaria bolanderi*

  6' Calyx radial, 5 or 10 almost equal lobes or teeth.

    • Calyx with 10 teeth, tips recurved .... *Marrubium vulgare*

    • Calyx with 5 sharp lobes. Corolla hairy inside, also usually outside (2 spp) .... *Stachys*

    • Calyx 10-veined, 5 not-quite-equal lobes. Corolla without upper lip, lower lip flat with 5 lobes, middle largest (2 spp) .... *Teucrium*

Hyptis—This large genus of about 350 species is widespread in warm temperate and tropical climates of the Americas, but California has just a single species. The Greek genus name means that the lower lip of the corolla has an enlarged central lobe with a pouch that is "turned back" (*hyptis*) or reflexed.

Hyptis emoryi (**desert lavender**) is a lanky, slender-branched shrub up to 3 m tall that frequents sandy washes and canyons of the Sonoran and southern Mojave Deserts under 1,000 m. Its flower clusters can be quite dense, but they are such a pale shade of lavender, and the foliage so gray a green, that the whole plant almost disappears into the landscape. However, the shrubs are highly noticeable to bees, which become

Hyptis emoryi (**desert lavender**)

so numerous as to form a haze around flowering branches. The light coloring of young twigs and leaves comes from a dense mat of hairs that are branched in a starlike radiating pattern. This same hairiness covers the calyx so densely that it is sometimes removed by birds for use as nest lining. Leaves are a wide oval, 1–2 cm long, with scalloped edges. Major W. H. Emory (1812–1887) directed the U.S.-Mexican Boundary Survey. The common name reference to lavender comes from the scent of the flowers as well as their color.

Salazaria (**bladder sage**)—This single-species genus is named for Don José Salazar, a Mexican commissioner and astronomer on the U.S.-Mexican Boundary Survey.

Salazaria mexicana (**bladder sage**) means *Salazaria* "of Mexico," referring to both the man the genus name honors and the fact that the plant is widespread in northern Mexico. This shrub, about 1 m tall, is highly distinctive for its densely branched rounded shape and for its greatly enlarged calyces

resembling bladders. Small branches spread widely and are spine-tipped. Each pair of leaves near the branch tips has its own flowers, one per leaf axil, quite neatly arranged. The upper lip of the corolla is white to light purple, and the lower lip is darker. As fruit develops, the calyx expands, becoming 1–2 cm in diameter with a translucent, papery texture. Back lighting in the late evening makes these dry shrubs glow with tiny illuminated paper bulbs. Bladder sage can be found in both deserts on gravel or sand, flats or slopes, under 1,800 m.

Salvia (**sage**)—With some 900 species worldwide, this large genus is well known. In part, its familiarity is due to various

Salazaria mexicana (bladder sage)

medicinal uses, recognized by the genus name meaning "to save." In addition, *Salvia* seeds are edible, and sage honey is a traditional bee product. *Salvia* may be annuals, perennials, or shrubs, usually with oval leaves that may be toothed, lobed, or even spine-tipped. Generally, both calyx and corolla are two-lipped, and the lower corolla lip often has an enlarged and elaborate central lobe. Each of the two fertile stamens has either one or two anther sacs; if there are two, one is larger than the other, and both are connected by a thread.

Salvia carduacea (**thistle sage**) has a Latin specific epithet meaning "thistle" (*cardu*), and indeed, the plant displays thistlelike features, including spine-

Salvia carduacea (thistle sage)

tipped, irregularly lobed, wavy-margined leaves. The calyx lobes are also spine-tipped. Corolla lobes are elaborately fringed and usually lavender to blue, but occasionally they are pure white. Stamens extend well out of the corolla tube and have bright red anthers. Each flower cluster grows at a leafing point that is encrusted with dense white woolly hairs. Thistle sage is a common annual in the western deserts under 1,400 m. Sometimes it forms large stands on sandy or gravelly slopes or flats where seeds have accumulated.

Salvia columbariae (chia) is another common annual, growing almost all over the state. *Columbariae* means that it "resembles *Columbaria*," or *Scabiosa* in the Teasel Family (Dipsacaceae), though this is not very helpful. Chia is a Mexican name, and Native Americans used chia seeds extensively in their diets. Plants grow quite reliably in dry sandy

Salvia columbariae (chia), Joshua Tree National Park, near the south entrance

Salvia funerea (Death Valley sage)

soils and disturbed areas under about 1,200 m, although occasionally they exist at higher elevations, where they are of smaller stature. Leaves, which grow mainly at the base of the plant, may be up to 10 cm long. They are pinnately lobed or dissected and irregularly rounded at the tips. Flowers are in very tight, perfect spheres no bigger than a ping-pong ball; a few of these may be spaced up each stem, with the smallest balls at the top. The calyces are about 1 cm long and partly dark purple. They dominate the smaller blue, barely noticeable corollas, which soon fall.

Salvia funerea (Death Valley sage) grows only in Death Valley and nearby mountains up to about 360 m—hence its specific epithet, meaning "having to do with a funeral" (*funerea*). The

Salvia mohavensis (Mojave sage)

lower end of Titus Canyon is a good place to find it. It is a compact, densely branched shrub about 1 m tall and almost silvery light gray-green. It is extremely woolly all over, and the leaves have several spines and wavy edges like small holly leaves. Corollas are violet to blue and about 1.5 cm long. When blooming profusely the flowers are showy, but the plant is interesting anyway for its Christmas leaves and snowy wool.

Salvia mohavensis (**Mojave sage**) is from mountainous portions of the Mojave Desert, but at elevations of only 300–1,500 m. Corollas are sky blue, with a long (to 2 cm) tube and lobes that are barely bilateral. Two stamens and a forked style are about twice the length of the corolla tube. Flower clusters usually form just at the tips of branches.

Examples and Uses

Here is a whole menu based on Lamiaceae! Start with mint (*Mentha*) juleps, move to fresh mozzarella and tomato salad with whole basil (*Ocimum*) leaves. Enjoy rosemary (*Rosmarinus*) rubbed leg of lamb accompanied by thyme (*Thymus*) flavored beans and tabouli with oregano (*Origanum*). For desert indulge yourself with peppermint (*Mentha*) sorbet alongside lavender (*Lavendula*) scented pound cake, and a cup of Oswego tea (*Monarda*) sweetened by sage (*Salvia*) honey. If this seems a bit much, consider that we left out marjoram (*Origanum*) and savory (*Satureja*).

Lavender is common in perfumes and sachets, and it is also lovely growing in the garden—or in the south of France, in large fields. Pennyroyal (*Hedeoma*) is another common perfume plant, and in Southeast Asia various species of *Pogostemon* are grown for the perfume industry.

There may be a teak bench in your herb garden, and this too is in the Mint Family, genus *Tectona*. Surely you have ornamentals from the family, such as *Cedronella* and *Coleus*. Cat lovers should be growing *Nepeta cataria*—catnip, highly prized by domestic felines. It makes a great exercise toy when dried and sewn into a cloth mouse. If catnip is growing in your garden, the neighborhood kitties will surely roll in it, so plant it where you can watch the show.

Hesperocallis undulata (desert lily), Sonoran Desert, Borrego Valley.

LILIACEAE

LILY FAMILY

What do Easter lilies, asparagus, and onions have in common? These and many other well-known plants belong to the extremely diverse family Liliaceae. Some are edible, some are poisonous, and nearly all are beautiful and easily spotted. Deserts have a few of the most spectacular representatives, such as Joshua trees and agaves.

Lilium, the Greek word for "lily," is the root of the name of the family, of the order, and of the subclass of monocots that have showy flowers with almost equal-sized petals and sepals. **Monocots** form a huge group of flowering plants that have a single cotyledon, or seed-leaf: the leaflike structure that appears immediately following germination of a seed. Other monocots include orchids, grasses, and palm trees.

In some treatments, Liliaceae is divided into 10, or even 40, families, so you may find family names such as Agavaceae and Asparagaceae. This book, however, follows *The Jepson Manual* (1996) in treating major groups as genera rather than as separate families. In the near future, DNA sequencing and reevaluation of taxonomic data may lead to another attempt to reorganize Liliaceae into more coherent families.

Liliaceae Icon Features

- Three sepals and three very similar petals, making a total of six perianth parts
- Six stamens, often attached to perianth lobes
- Three stigma lobes, reflecting three chambers of the ovary

Distinctive Features of Liliaceae

Correct identification of Liliaceae members is facilitated by first being able to distinguish monocots from dicots. Since we do not usually see sprouting seeds and flowers at the same time, however, we must rely on other clues to monocots, including leaves with the larger veins parallel to one another and flower parts in threes.

In Liliaceae, petals and sepals are three each, for a total of six perianth parts in radial symmetry. You may see that the perianth parts are attached in two tight concentric circles, the inner one consisting of three petals, and the outer one, of three sepals, arranged alternately with the petals. Sepals usually are shaped and colored almost identically with petals. Sometimes the perianth has a tubular base with six lobes. Many desert genera have a white perianth, but shades of blue to lavender, and occasionally yellow to orange or even red, are found as well. Individual flowers are almost always at least 1 cm across and often occur in clusters.

Liliaceae flowers generally have six stamens that may or may not be attached to the perianth and may or may not include various appendages or sterile members (staminodes). There is one ovary, either superior or inferior, with three chambers, one style, and three stigmas or stigma lobes. In our desert lilies, only *Agave* species have an inferior ovary. Fruits usually develop three distinct bulges, making the location of chambers apparent, and as fruits mature and dry they split apart at

Free perianth parts
Sepals and petals are very similar and together are called the perianth.

— Sepal

— Petal

Death camas (*Zigadenus venenosas* var. *venenosas*)

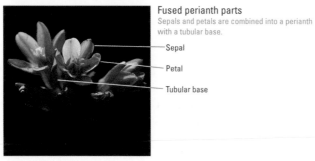

Fused perianth parts
Sepals and petals are combined into a perianth with a tubular base.

— Sepal

— Petal

— Tubular base

Blue dicks (*Dichelostemma capitatum*)

the tip, revealing the insides of chambers with their seeds.

Lily Family members are usually perennials with an expanded underground stem, such as a bulb, that sprouts a circular cluster of leaves spreading from ground level. In this situation, flowers often are borne on a leafless stem also arising from ground level. Other bulbs may sprout stems with small leaves arranged alternately or in whorls.

A few desert Lily Family members are tree- or shrublike, with long pointed leaves, usually looking like a compact spiraled cluster of dangerous sword blades growing near the base of the plant. A few species with elevated branches have their swords in clusters near the branch tips, and occasionally leaves are sized more like daggers.

Lily Family (Liliaceae) flowers often have six separate perianth parts, but sometimes they are fused at the base, forming a tube with six lobes.

Similar Families

The all-inclusive definition of Liliaceae adopted here leaves little room for similar families. Almost everything that looks like a lily is a lily.

Iris Family members (Iridaceae) are sometimes mistaken for lilies, but usually irises have inferior ovaries and sepals that are showy but very differently shaped from petals. Although most irises do not grow in deserts, the genus *Sisyrinchium* has three species present in our deserts, and their sepals and petals are very similar, as in Liliaceae. These desert *Sisyrinchium* grow in alkaline meadows or wet areas and have pale blue to deep blue-violet perianths. There are usually several stems in a cluster with narrow grasslike basal leaves, and often a few leaves along the stems. The habitat and color together distinguish them from desert lilies.

Orchids (Orchidaceae) are the only other family of monocots with showy perennial desert flowers. Although Orchidaceae is the world's largest plant family, only two species of orchid inhabit the desert. Stream orchid, *Epipactus gigantea*, grows only in cool streams and wet areas, where it can pretend it is not in the desert. It is immediately recognizable as an orchid by the strong bilateral symmetry of flowers and its unusually shaped perianth parts. White-flowered bog orchid, *Platanthera leucostachys*, grows only in wet places high in the Panamint Mountains. Its small flowers have clear bilateral symmetry, unlike those of Liliaceae members.

Family Size and Distribution

Liliaceae contains about 300 genera and 4,600 species worldwide. Species are concentrated in temperate and subtropical climates, and many prefer moist, lightly shaded locations—not the usual condition of deserts. Many species have been spread by cultivation for their beautiful flowers. Species, like tulips and daffodils, that multiply from bulbs are simple to propagate true to color and easy to establish in new locations. Plant breeders growing new plants from seed for the occasional spectacular result are responsible for the hundreds of varieties available to us through catalogs.

California has 40 genera of Liliaceae, six of which are represented entirely by nonnative species, like *Aloe* and *Asparagus*, that have escaped from cultivation. Eleven of California's genera, all composed primarily or entirely of native species, have desert representatives. A few genera are restricted mainly to desert locations.

The Joshua tree, *Yucca brevifolia*, though widespread throughout the Mojave Desert, seldom extends either north or south from there. It is sometimes called an indicator species for the Mojave Desert, and its presence has been used to map boundaries of this region. Only a few other species of Liliaceae have as broad a desert distribution as Joshua trees. Many species present in California's deserts are found only in desert mountains, and even there their range is frequently restricted.

California Desert Genera and Species

The easiest Liliaceae members to spot are the tall, shrubby or treelike forms with evergreen, swordlike leaves. In spring and early summer they have huge clusters of white or yellow flowers that are utterly breathtaking. Other Liliaceae genera, the perennials, sprout each year from underground bulbs or similar structures. Their flowers are often recognizable as resembling the tulips, lilies, or onions that grow in gardens.

SIMPLIFIED KEY TO DESERT GENERA OF LILIACEAE

Most genera of Liliaceae found in California's deserts can be distinguished by growth form, including how the leaves are arranged and shaped. Flower arrangements are also important, and in a few cases you will need to look at perianth parts or stamens. In general, the genera are distinctly different and easy to learn.

1. Plants large and tree- or shrublike. Leaves evergreen, long and pointed, arranged in large spiraling clusters. Flowers white or yellow, arranged in long formations rising from the centers of leaf clusters.
    2. Leaves stiff but fleshy, with marginal teeth. Flowers yellow or creamy white and growing in clusters on a tall flowering stalk (2 spp) .... *Agave*
    2' Leaves stiff or not, never fleshy or toothed but often with shredding marginal fibers. Stems often with woody branches. Flowers white, sometimes with reddish or greenish markings, especially on buds.
        • Leaves very stiff and spine tipped. Flowers larger than 3 cm (4 spp) .... *Yucca*
        • Leaves somewhat flexible, not spine tipped. Flowers less than 1 cm (2 spp) .... *Nolina*
1' Plants perennial. Leaves seasonal and withering. Flower color and arrangement various.
    3. Leaves entirely basal. Flowers in bracted clusters on a leafless stem arising from ground level.
        4. Perianth parts separate, not forming an obvious tube at the base.
            • Plants with onion odor (7 spp) .... *Allium*

• Plants without onion odor (2 spp) .... *Muilla*

4' Perianth parts fused into a tube at the base.

　• Stamens of 2 distinct sizes, 3 much larger than the other 3 .... ***Dichelostemma capitatum***

　• Stamens equal in size with filaments fused into a tube .... *Androstephium breviflorum*

3' Leaves usually basal and along stems but generally reduced upward. Flowers not in distinct bracted clusters just at the tip of a leafless stem that arises from ground level.

　5. Basal leaves 0–2 and withered at flowering. Perianth white or colored.

　　• Flowers nodding (pointing downward). Perianth purplish brown with yellow to white mottling throughout. Leaves along stem in 2s or 3s .... *Fritillaria autopurpurea*

　　• Flowers upright. Perianth resembling a tulip, creamy white, yellow, orange, red, or lavender. Basal leaves 1–2. Stem with zero to few leaves or leaflike bracts (4 spp) .... ***Calochortus***

　5' Basal leaves several to many and green at flowering. Perianth white.

　　• Flowers many, spreading from stem. Perianth 6–8 mm (2 spp) .... *Zigadenus*

　　• Flowers few to many, spreading from stem. Perianth 40–60 mm .... ***Hesperocallis undulata***

Agave (**century plant, agave**)—California has only three species of *Agave*, but there are around 300 in other parts of the Americas, including the tropics. The Greek word *agave*, meaning "noble," refers to the large size of plants and flower stalks. Large clusters of sharply pointed leaves are shrublike, with several clusters grouped together and growing for many years (but not a century) without flowering. Leaves are up to 40 cm long, like short swords, but they are thick and fleshy and both edges have teeth. Finally, when conditions are just right, a large cluster in the group will send up a flower stalk that looks, at first, like a gigantic asparagus. The top of the stalk bursts into flower, sets seed, and then dies along with the cluster of leaves from which it grew.

The long period of growth prior to flowering and the finality of flowering mean that successful germination of seeds and growth of new plants need to be fairly dependable. Plants have evolved to flower only during favorable conditions so they are likely to complete the process of making seeds. Seeds released from the high position of flowers have a good opportunity to disperse and a fair chance of finding suitable conditions for bringing on a new generation. Cahuilla Indians stopped the process of seed making by prying out the large sweet buds of flower stalks and roasting them for food.

Agave deserti (**desert agave**) grows on rocky slopes and washes of the Sonoran Desert and southern desert mountains under 1,500 m. Its sharply pointed, gray-green leaves have widely spaced marginal teeth. Clusters of yellow flowers grow on

Agave deserti (**desert agave**)

short spreading branches near the top of the flowering stalk, way above a person's reach. The only other member of the genus in our deserts is Utah agave, *Agave utahensis*, found just in the northeastern desert. It does not have a branching flower stalk. Cream-colored flowers grow in small clusters right along the main stalk.

Allium (**onion, garlic**)—The Latin word for garlic provides the genus name for this large group of plants. There are 500 species worldwide, some of which are probably in your kitchen. Taste and odor are reliable ways of identifying this genus, and our seven desert species are no exception. Other features include hollow cylindrical leaves (nondesert species may have flat or channeled leaves) and flowers in a distinct cluster on leafless stems arising from ground level. Onions usually have two to four bracts—dry, papery, leaflike structures—just below the flower cluster.

Identification of *Allium* species generally requires microscopic inspection of bulb coatings, but these are not easy to obtain or wise

Allium fimbriatum (**fringed onion**)

to dig because, of course, it terminates the plant. The paucity of desert species generally makes the task of identification a bit easier, but there still are some difficult distinctions.

Allium fimbriatum (**fringed onion**) grows on dry slopes of the western deserts. Its bare stem is 10–37 cm tall, and its cylindric leaf is 1.5–2 times the length of the stem. The superior ovary has six pointed and toothed projections on top, accounting for the common name of "fringed." The six perianth parts are usually pink or lavender but may be white, and there are 12 or more flowers per cluster.

Calochortus (**mariposa lily**)—This special Liliaceae genus, called "beautiful" (*calo-*) "grass" (*chortus*) in Greek, comprises 65 species, of which California has 43. The flowers, slightly resembling tulips, are certainly beautiful, so the Greek name makes sense. The common name, from the Spanish *mariposa*, meaning "butterfly," refers to the lovely coloring and shape of the flower petals. The bulbs of many species were eaten by Native Americans. The four species that occur in California's deserts include some of the most spectacular ones, and if you were hungry enough they would be easy to spot. Getting to the bulb might be another matter.

Calochortus flexuosus (**straggling mariposa**)

Calochortus flexuosus (**straggling mariposa**) has a "flexing" or "winding" leafless stem, often "straggling" up through low shrubs of the eastern deserts. One or two leaves, usually dried by the time of flowering, grow at the base of the stem, but bracts at branching points on the stem resemble leaves. The tuliplike

Calochortus kennedyi var. *kennedyi* (desert mariposa)

Calochortus kennedyi var. *munzii* (desert mariposa)

Calochortus panamintensis (Panamint mariposa lily)

perianth is made up of three wide lavender petals with yellow bands and three narrow sepals that are usually tinged green.

Calochortus kennedyi (**desert mariposa**) is a car-stopper (even at 70 mph), with brilliant coloring that varies from intense yellow to radiant red. Most often petals are electric orange with an almost black spot (the nectary, or nectar-producing gland) at the base of each petal. Sepals are narrow and greenish. Up to six flowers may grow on a single stem, standing straight up to boldly announce their location. It can be very exciting to find a population, since they do not flower every year and are seldom abundant. The species was named for William L. Kennedy (c. 1827), who collected specimens for the botanist who described it.

Calochortus panamintensis (**Panamint mariposa lily**) is a rare species found only in the Panamint Mountains at 2,500–3,200 m. Its pale creamy coloring is more typical of many

Calochortus species than is the brilliant coloring of *Calochortus kennedyi*. Distinguishing features include the fairly narrow (not overlapping) petals, whose only markings are a dark red or purple nectary spot surrounded by a fringe and a green vertical line on the back. Sepals are colored similarly to petals but are much smaller.

Dichelostemma—This genus has only five species, most of which grow in northern California and points north, but one species with a very broad range dips into the deserts. The growth form of *Dichelostemma* resembles that of *Allium*, with a bracted cluster of flowers on the tip of a bare stem and just a few basal leaves, but *Dichelostemma* do not have an onion odor. The Greek name *Dichelostemma* (*dichelo-* meaning "cloven hoof" and *stemma* meaning "garland" or "crown") is usually translated "toothed crown" and refers to pointed appendages on the stamens. The perianth of *Dichelostemma* is fused at the base into a tube, with six lobes spreading above where anthers are attached and the "toothed crown" is evident.

Dichelostemma capitatum (blue dicks)

Dichelostemma capitatum (blue dicks) has a blue to pinkish or occasionally white perianth. The term "dicks" is short for *Dichelostemma*. *Capitatum* is Latin referring to the "headlike" rounded cluster of flowers. Blue dicks has basal leaves that are just slightly keeled or ridged on the underside. Flowers have three large stamens and three much smaller ones. Blue dicks are common on desert slopes up to 2,300 m, and sometimes grow in small but dense colonies.

Hesperocallis—This is a single-species genus with a Greek name meaning "western," as in setting sun, or "evening" (*hespero-*) "beauty" (*callis*).

Hesperocallis undulata (desert lily) resembles Easter lilies, with rather large (4.5–6 cm) fragrant white flowers, tubular near the base but with

Hesperocallis undulata (desert lily)

spreading lobes that have a greenish stripe on the back. Long, narrow leaves are mostly basal, and a stem that can grow to over 1 m bears some smaller leaves and several to many flowers, each with a bract. In good years stems may be branched and flowers numerous. The specific epithet, *undulata*, comes from the "wavy" or "undulating" margins of the leaves. The bulb is edible but deep in sand: definitely not worth the effort of retrieval, since it tastes like garlic and besides, its harvest would eliminate the plant. These beautiful lilies grow only under 800 m in the deserts of California and western Arizona.

Nolina (beargrass)—The eighteenth-century French agricultural writer P. C. Nolin is honored in the genus name *Nolina*. There are about 25 species in Mexico and the southern United States, three of them occurring in California. Two are found in our desert mountains and the Sonoran Desert.

Nolina parryi (Parry nolina), Mojave Desert, Kingston Range

These are tall (2 m) treelike plants with huge clusters of long thin leaves, sometimes 200 in a single cluster. Beargrass probably refers to the robust grassy look of leaves that are somewhat flexible and are not sharply pointed. The flowering stems can be 3 m tall with hundreds of small (under 1 cm) white flowers. Usually, flowers of a single sex occur on a plant, but often along with some bisexual flowers.

Nolina parryi (Parry nolina) was first collected by Dr. C. C. Parry (1823–1890) near Mount San Jacinto. Finding these huge plants would have been easy compared to collecting them! They can be spotted from a mile away, especially if there is a plume of white flowers rising above the clusters of leaves. *Nolina parryi* may have a few branches, each with a huge cluster of leaves. Leaf margins do not shred and become fibrous at maturity as in the other desert species, *Nolina bigelovii*. *Nolina parryi* grows at elevations of 900–2,100 m.

Yucca (Spanish bayonet, yucca)—The name yuca (with one *c*) is the Haitian name for cassava, a totally different plant in the Spurge Family (Euphorbiaceae). For some reason, *Yucca* was named after yuca. The name Spanish bayonette reflects the fact that the leaves of most *Yucca* species are dangerous weapons—rigid and sharply pointed. Although the usual arrangement of leaves in huge clusters is

similar to *Nolina,* *Yucca* leaves are far more rigid than their more grasslike cousins. Flowering stalks of *Yucca* sometimes resemble *Nolina,* with long plumes of white flowers, but *Yucca* flowers are larger (over 3 cm) and they are always bisexual. *Yucca* are night blooming and attract special species of moths that generously pollinate flowers and then lay their eggs in the flower's ovary. Moth larvae develop as fruits mature and eat a few seeds before they munch through the fruit wall and fall to the ground to over-winter as cocoons. Moths emerge

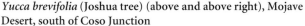

Yucca brevifolia (Joshua tree) (above and above right), Mojave Desert, south of Coso Junction

just in time to pollinate the next season's flowers. There are about 40 species of *Yucca,* found especially in the dry southwestern portions of North America. All four species found in California are restricted to our deserts and nearby semiarid zones.

Yucca brevifolia (**Joshua tree**) is unusual and distinctive, even among its closest relatives. Growing to a height of up to 15 m, it is often heavily branched. The upturned branches reminded Mormon settlers of the prophet Joshua, raising his arms to heaven, and so the tree obtained its common name. Clusters of leaves grow at the tips of branches, leaving the trunk bare and woody looking. Leaves are only 20–35 cm long, unusually "short" (*brevi-*) "leaved" (*folia*) for the genus. Dense spikes of flowers form at the ends of several branches in years when conditions are favorable. Flowering causes growth of tissue at the tip of the branch to cease, leading to formation of additional branches that can flower in future years. In addition to growing from seed, Joshua trees spread by sprouts from long underground runners. The Joshua tree, for which the national park is named, grows over huge expanses of the Mojave Desert and essentially nowhere else, so it is considered an indicator species for Mojave Desert. As might be expected for so imposing a tree, many species of birds find it useful for nest sites, and you might find it the only source of shade for your picnic lunch.

Yucca schidigera (Mojave yucca) is usually shorter and less branched than *Yucca brevifolia*, but its leaves are much longer (up to 150 cm) and it can be a very impressive plant. Large flower stalks form at the ends of branches, and flowers are large (3–5 cm). It is common in the southern Mojave Desert and in northwestern portions of the Sonoran Desert. *Schidigera* sounds like someone's name, but it is probably a composite Greek/Latin word meaning that "little splinters" (*schidi-*) are "borne" or "carried" (*-gera*) on the margins of the leaves, or it may combine Greek words to mean "splintering" (*schidi-*) in "old age" (*-gera*). In either case the reference is to fibrous shredding along leaf margins.

Yucca schidigera (Mojave yucca), Joshua Tree National Park

Zigadenus (death camus)—This genus has about 15 species in temperate North America and Asia. As the name implies, most species are highly toxic and sometimes kill grazing live-stock. The Greek word *zyg* means "yoke," as in a pair of yoked oxen, and refers to pairs of "glands" (*aden*) at the base of perianth parts. Plants grow from a bulb and have long, narrow, mostly basal leaves, but a few, reduced in size, grow up the stem. Flowers are white to yellowish, and glands at the base of each perianth segment are helpful in identification of genus and species.

Zigadenus venenosus (death camus) likes moist places such as seeps or meadows of desert mountains. Its Latin name means "full of" (*-osus*) "poison" (*venen-*). Its flower stalk grows to only about

Zigadenus venenosus (death camus)

20 cm, and its flowers are only about 6 mm long, with the inner perianth lobes slightly longer than the outer. Stamens may extend beyond the perianth, which is white with yellowish green glands.

Examples and Uses

The Liliaceae members best known for their flavor are in the genus *Allium*, which includes onions (*Allium cepa*), garlic (*Allium sativum*), and leeks (*Allium porrum*). Together with similar genera, like *Agapanthus* (lily-of-the-Nile), *Allium* is frequently assigned to the family Alliaceae.

Asparagus (*Asparagus offinalis*) is another well-known edible member of Liliaceae, and once again, it is often placed in a separate family, Asparagaceae, along with close relatives such as *Polygonatum* (Solomon's seal) and *Convallaria* (lily-of-the-valley).

Another familiar group of Liliaceae is characterized by the genus *Narcissus* (daffodils and jonquils), *Clivia* (African-lily), *Galanthus* (snowdrops), and several other ornamentals often referred to as amaryllis, and frequently assigned to the family Amaryllidaceae. These species are particularly diverse in South Africa, the Mediterranean, and the Andes of South America, so they are known to North Americans primarily as gardening plants.

Still more familiar ornamentals are in the genera *Lilium* (lilies), *Tulipa* (tulips), *Fritillaria* (fritillarias), and *Erythronium* (dog-toothed violets). Together with *Calochortus* (mariposa lily), which is not readily cultivated, this group is at the heart of Liliaceae, and in most treatments is kept together in the family Liliaceae even when others are split into separate families. Many species native to North America are included in this group. Tulips, however, originated in Turkey and were introduced to western Europe in the sixteenth century, where they exploded in popularity. Since then, tulips have been cultivated by the Dutch to the point of becoming a national symbol.

Agave and *Yucca* are a useful as well as ornamental group of Liliaceae often placed in the family Agavaceae. Tequila and mescal both are made from the sugary fermented sap of *Agave*. Rope and cord for various weavings and stitchings are made from leaf fibers of both *Agave* and *Yucca*. Dense plantings sometimes are used for fencing—and indeed, without a machete you would be hard pressed to cross.

Eucnide urens (sting nettle), Death Valley National Park, mouth of Titus Canyon

LOASACEAE

LOASA OR BLAZING STAR FAMILY

Intense yellow flowers on a brittle annual (sometimes perennial) are typical of *Mentzelia*, the largest genus of Loasaceae in our deserts. Subshrubs, usually with white or cream flowers, are the other common form you will see. Some members of this family announce themselves with stinging or barbed hairs that discourage too close a look. Stiff barbed hairs may keep some gnawing insects at bay, but will more likely act like velcro to attach bits of the brittle twigs and leaves to your shirt. The same mechanism is important in dispersing fruits full of seeds carried by unwitting animals doing the work that the plants cannot do themselves.

The word *Loasa,* root for the family name, is probably a native South American name for these and similar plants. Blazing star is a direct reference to the bright, reflective, five-pointed flowers of typical members of the family.

Loasaceae Icon Features

- Fruits formed from an inferior ovary and usually covered by stiff hairs
- Five sepals at the top of the ovary remaining attached to the fruit
- Single style remaining attached to the fruit after petals have fallen

Distinctive Features of Loasaceae

Annuals, perennials, or subshrubs, with sharp or sandpaperlike hairs on leaves and stems, are typical of the family. Some plants have stinging or barbed hairs. You may also notice a whitish cast to the stiff stems, which is common but not universal. Usually, there are five petals (one species has eight) that may either be separate or fused to each other or to the base of stamens. White and yellow with a glossy polish are the most common flower colors. Size varies from tiny to showy.

The ovary is inferior, often looking like a fat, cylindrical section of stem, with sepals and style still attached to the top end after petals have fallen. There are usually many stamens, but one genus in California (*Petalonyx*) has just five. Filaments may be flattened or originate in clusters. Occasionally, some of the outermost filaments grow wide like petals and lack their pollen-producing anthers, in which case they are called **staminodes**. The pistil has only one style, although the stigma may have three or five lobes.

Similar Families

An inferior ovary is somewhat unusual, but for positive identification this feature must be combined with a single style, distinctive hairs, five petals (with one exception, a consistent feature of desert species), and numerous stamens (except for *Petalonyx*). When counting petals, be careful not to include expanded staminodes. Although this may sound complicated, the five-pointed blazing star flowers are not easily confused with other inferior-ovary families having a single style. In the desert, these are primarily Evening Primrose Family (Onagraceae), which usually has four petals and four or eight stamens, and Cactus Family (Cactaceae), which has numerous petals and other prominent features such as dangerous spines and succulent stems.

Family Size and Distribution

Loasaceae comprises 15 genera and about 200 species, of which all but two unusual members of the genus *Fissenia* are native to the Western Hemisphere. Family members are most abundant in South America, where they are found in both tropical and temperate climates and wet and dry habitats. The largest genus, *Loasa*, with about 75 species, is not present in California.

Several genera occur in southwestern deserts of North America, but there are only three in California, all well represented in California's deserts. All species of Loasaceae found in California are natives, and they grow in hot, dry places with sandy or rocky soils or even virtually no soil. There is a species for almost any elevation, although they tend to be more abundant and diverse at lower elevations.

California Desert Genera and Species

The most noticeable desert species have conspicuous (up to 4 cm across) bright yellow flowers growing on sparsely leaved and brittle-looking whitish stems. The most common species have flowers about 1 cm across, and so closely resemble one another that a microscope and mature fruits are necessary for accurate identification. Several less common species have features that allow them to be readily identified.

SIMPLIFIED KEY TO DESERT GENERA OF LOASACEAE

1. Stamens 5. Petals white or cream and 1 cm or less long, edges sometimes fused near the middle. Hairs generally sandpapery (3 spp) .... *Petalonyx*

1' Stamens many. Petals generally yellow, rarely greenish, if cream then larger than 1 cm. Hairs various.

 • Annual with petals greenish or subshrub with petals cream and 3–5 cm. Hairs long and stinging or barbed. Leaves almost round and toothed to lobed (2 spp) .... *Eucnide*

 • Annuals and perennials. Petals generally yellow, some pale, some almost orange or with orange markings. Hairs generally sandpapery. Leaves generally longer than wide, with or without teeth and deep lobes (20 spp) .... *Mentzelia*

Eucnide (**rock nettle**)—This genus has eight species, all growing in the southwestern United States and northern Mexico. The two California species are both desert inhabitants. The Greek words *eu-* (good or true) and *-cnide* (nettle) tell us this is a serious nettle, prepared to sting.

Eucnide urens (**sting nettle**) is a subshrub with broad leaves that look soft and fuzzy because of all the hairs. As the common name implies, you will be sorry if you touch it. Even the specific epithet, *urens,* warns that this plant is "stinging" or "burning." (Note that the same common name applies also to plants in the Nettle Family, Urticaceae, that have stinging hairs.) *Eucnide urens* is common in

Death Valley in rock crevices of canyons and at the base of canyon walls. In full flower it seems an unlikely inhabitant of the rocky landscape, covered as it is with broad (3–5 cm) creamy flowers. It grows in both the Mojave and Sonoran Deserts under 1,400 m but is most common in eastern portions of California.

Mentzelia (blazing star)—The German botanist Christian Mentzel (1622–1701) provided the root for this genus name. The common name, blazing star, refers to the five free petals (though one species, *Mentzelia reflexa*, has eight petals) and their bright yellow color (usually). There also are many stamens, and sometimes also staminodes, giving some species a

Eucnide urens (sting nettle)

doubled look similar to cultivars that have been bred for an extra set of petals. Although the fruits are basically cylindrical, the many variations on this shape, and on the shapes and surface textures of the seeds, are important in identification of species.

Mentzelia species in California's deserts may be annuals, biennials, or perennials, and they almost always have rough or barbed hairs, noticeable if you feel the leaves, stem, or fruit. Most species have

a group of petioled leaves growing close to the ground and smaller leaves without petioles growing along the stem, often widely spaced, at branching points.

Of the 28 species of *Mentzelia* in California, 20 are found in the deserts, and most are hard to key. The following descriptions highlight distinctive features, but a good species key, a microscope, and lots of patience will generally be necessary for positive identification of many species.

Mentzelia albicaulis (small-flowered blazing star) is an annual with a "white" (*albi-*) "stem" (*caulis*), generally branched and growing up to about 40 cm. Yellow flowers are small (petals 2–7 mm) and sometimes have a faint darker yellow spot

Mentzelia albicaulis (small-flowered blazing star)

Mentzelia eremophila (desert blazing star)

at the base. Fruit is usually straight but may be curved into nearly a semicircle. A very similar species is *Mentzelia obscura*, which overlaps in size but is generally a bit larger and has a stem that is less white than that of *Mentzelia albicaulis*. Other white-stemmed species either are perennials, usually with peeling white stems, or have flowers of a different size or color. In keys, *Mentzelia albicaulis* is distinguished from *Mentzelia obscura* by having a more angular seed surface. Both seed surfaces are covered with tiny warts so they look well pebbled, but seeds of *Mentzelia albicaulis* have both more pointed warts and more acute corners to the overall shape of the seed. *Mentzelia albicaulis* grows throughout the deserts at elevations from 500 to 2,300 m.

Mentzelia eremophila (**desert blazing star**) is an annual 10–40 cm tall with thin, spreading branches. Flowers generally are few but large, with bright yellow petals about 2 cm long and broad. Stamens are only 3–10 mm. This species is found in washes and along roadsides of the southern Mojave Desert. Its Greek specific epithet means "loving" (*-phila*) to be a "desert dweller" or "lonely" (*eremo-*).

Mentzelia involucrata (**sand blazing star**) is easy to recognize because of its large (up to 6 cm) light yellow petals with distinctive orange veins. Fruits are long (14–22 mm), fat (5–10 mm), and straight. The specific epithet, *involucrata*, refers to the prominent "involucre," which consists of four or five toothed or lobed bracts below each flower. The bracts are similar to small leaves but are green just on their edges and otherwise are whitish or dried looking. *Mentzelia involucrata*, an annual often growing in small patches, can be found in hot sandy washes below 900 m.

Mentzelia laevicaulis (**giant blazing star**) is a large and spectacular perennial with a distinctly erect main stem and many upper branches, growing to almost 1 m in good conditions such as road cuts. It grows all over California except for the Sonoran Desert and the

Mentzelia involucrata (sand blazing star)

Mentzelia laevicaulis (giant blazing star)

Mentzelia multiflora ssp. *longiloba* (Adonis blazing star)

Central Valley, at elevations below 2,700 m. The specific epithet means "smooth" (*laevi-*) "stem" (*caulis*). *Mentzelia laevicaulis* has very long (15–46 mm) sepals and even longer (30–80 mm) yellow petals, with lots of long (15–55 mm) stamens. Outer stamens may be a tiny bit wider than the inner ones, but they are not staminodes. In full bloom, this plant is a show stopper. After a couple of months without rain, however, it becomes a straw-colored skeleton, with its nutrients loaded into the root stock waiting to sprout in the next rainy season.

Mentzelia multiflora ssp. *longiloba* (**Adonis blazing star**) is a "many" (*multi-*) "flowered" (*flora*) perennial when it grows to its full size of up to 1 m. The five petals are 1–2 cm long and rounded at the tip. Five broad staminodes fill out flowers, giving them the look of having 10 petals. Fruits are fat for the genus—up to 1 cm wide—and seeds are large (3–4 mm long) with a distinctive wing. This is a low-elevation (under 700 m) species, choosing sandy habitats of the Sonoran or southern Mojave Deserts.

Petalonyx (**sandpaper plant**)—The Greek word *onyx*, meaning "claw," refers to the L-shaped petals of this genus. Petals are sometimes fused together in the middle near their bends, and stamens then grow out through the holes so that they seem to be attached outside the petals rather than inside. Flowers are small (about 1 cm or less), so this feature requires close observation to discern. Three species of *Petalonyx* occur in California, all subshrubs growing in the deserts, and almost all with stiff barbed hairs on the leaves, feeling like sandpaper.

Petalonyx nitidus (**smooth sandpaper plant, shiny-leaved sandpaper plant**) grows to 50 cm in height and is a very leafy, heavily branched, stiff, and sandpapery (not smooth) subshrub. The specific epithet refers to "shining" or "handsome" (*nitidus*) leaves, which are up to 4 cm long and broad. Flowers are cream-colored with petals 5–11 mm long, most of which is claw, below the spreading lobes. Large (3–5 cm) clusters of flowers grow on the ends of numerous branches. This species grows in the sandy or

Petalonyx nitidus (smooth sandpaper plant)

rocky canyons of the desert mountains.

Petalonyx thurberi (**Thurber sandpaper plant**) can grow up to 1 m in height but has very small (about 5 mm) heart-shaped leaves on the branches that bear flowers, with bigger leaves below, so it sometimes appears open or even frail. Flowers are small (2.5–7 mm) and cream-colored, with petals fused at the bending point below the spreading and expanded tips. Stamens, which are 4–10 mm long, extend out

Petalonyx thruberi (Thurber sandpaper plant), Mojave Desert, east of Granite Pass

through the holes between petals, surpassing the petals in length. Stems and leaves have the characteristic sandpaper feel in the subspecies *thurberi*, which grows throughout the deserts, but the rare subspecies *gilmanii*, known as Death Valley sandpaper plant, has hairs that are soft. It is confined to the northern Mojave Desert.

Examples and Uses

You may not be aware of any examples of this family unless you have seen the few species of *Mentzelia* or *Blumenbachia* that are used as ornamentals, or unless you have been victimized by sting nettle (*Eucnide urens*), which grows in washes and on rocky slopes of the low deserts. We are unaware of any other uses for Loasaceae, although sandpaper is a possible product of some.

Sphaeralcea ambigua var. *ambigua* (apricot mallow), Death Valley National Park, Last Chance Range

MALVACEAE

MALLOW FAMILY

Some of the most eye-popping orange and rose flowers of a good spring desert bloom arrive on members of the Mallow Family. It is hard to imagine they have anything to do with the marshmallows that campers cook on sticks, but they do (check out our section on Examples and Uses). Although desert species will not add to your late-night campfire experience, in the low evening light mounds of bright orange flowers may look like a blazing fire.

The Latin word *malva,* referring to the common mallow plants of the Mediterranean region, is the root for the family name Malvaceae. Mallow in an Anglicized version of the Latin.

Malvaceae Icon Features

- Stamens with fused filaments forming a tube
- Numerous anthers spreading from the filament tube
- Long style emerging above stamens and branched (usually) at the tip

Distinctive Features of Malvaceae

The infallible feature to look for in Malvaceae is numerous stamens with their filaments fused together forming a tube that surrounds the style portion of the pistil. Thus, male and female parts of the flower form a complex-looking central structure in the middle of five showy petals. Stamens are arranged in a circle surrounding the ovary, their filaments joined at the sides to form a long tube. Anthers protrude from the tube in various patterns. The style, extending above the ovary, is long and thin and surrounded by the filament tube. As the style exits the tip of the filament tube, it branches into several (often five) small stalks holding somewhat bulbous or linear stigmas. The five petals of Malvaceae flowers are not fused or attached to each other, but because they are all attached to the central filament tube, petals fall together with the tube.

Hibiscus, with its large, conspicuous arrangement of stamens and pistil, is probably the best-known member of the family. Other examples may be smaller or less obvious, but the basic pattern is consistent throughout the family. If in doubt, cut lengthwise through the center of the flower and look for the tube of filaments with the style loose inside it.

The ovary in Malvaceae is superior and five- (sometimes two-) to many-chambered. Usually, chambers are arranged in a ring around a central post. When mature, the chambers tend to break apart like the sections of chocolate oranges you might have gotten from your Dutch neighbors at Christmas time. If you need to identify Malvaceae genera using a key, it is important to have fruits; however, many desert species are easily recognized by color and form of the flower alone.

Vegetative parts of Malvaceae also are distinctive. Leaves are alternately arranged and have stipules, looking like small wings or points on either side of the petiole where it attaches to the stem. Leaf blades are palmately veined and usually lobed around the major veins. If you pick some flower stems you may notice the tough, fibrous inner bark and the sticky, mucilaginous sap. Looking more closely, note the numerous star-shaped hairs on the stem surface, and the calcium oxalate crystals in the tissue.

Similar Families

Although three other families bear many similarities to Malvaceae, there is nothing in the desert that will confuse you. Bombacaceae species are tropical trees (the kapok tree). Tiliaceae members, largely tropical, are represented by linden trees in North America and Europe but are not present in California outside of cultivation. Sterculiaceae, or Cacao Family, important in the tropics as the source of chocolate, has only one representative in California's deserts, *Ayenia compacta,* a rare shrub with tiny (3 mm) flowers. Its very peculiar petals rise on a slender claw and expand where they attach (more or less) to each of five anthers that spread from the filament tube. Although this flower is technically very similar to Malvaceae flowers, it is so unusual that you will know immediately it is not a mallow.

Family Size and Distribution

Malvaceae is found worldwide, especially in warm climates. There are roughly 100 genera and 2,000 species, but only 16 genera occur in California, five of which are represented solely by nonnative species. The family namesake genus, *Malva,* is one of those occurring only as a weed (or ornamental) in California, including our deserts. Its common name, cheeseweed, refers to the ovary segments arranged like wedges in a wheel of cheese.

Of the 11 genera with native California species, eight have representatives in the deserts. Native genera have an unusually high number of uncommon, rare, and endangered species, a feature consistent with their very limited distribution; many of these species are found only in localized areas that may be severely affected by development, grazing, water diversion, and other impacts to the habitat. Fortunately, a few of the most colorful desert species are fairly widespread.

California Desert Genera and Species

Desert species of Malvaceae (which, with a few notable exceptions, are generally uncommon) are annuals, perennials, subshrubs, or rarely shrubs, with colorful flowers of small to large size (up to 5 cm across). Leaves of desert Malvaceae members tend to be broad and hairy and to grow well up the stems. Often, flowers grow on pedicels from the leaf axils.

SIMPLIFIED KEY TO DESERT GENERA OF MALVACEAE

Although the family is easy to recognize, genera, and especially species, can be difficult. The patterns in which anthers extend from the tube of fused filaments are diagnostic of some genera. There may be a continuous pattern below the tip like a tiny bottlebrush, all anthers may be consolidated at the top of the column, or they may extend from the column in five radiating clusters. Sometimes leaf features are helpful in distinguishing a genus, although here, too, considerable variation within genera exists. Petal size and color are not very consistent within genera either, except where there are so few species that generalizations work. It is often helpful to have fruits so that the number of segments, number of seeds per segment, and features of the seeds can be evaluated. Overall, only a combination of features may lead to identification of a genus without the need to examine technical details.

1. Flowers single in leaf axils and large (about 4–5 cm across). Petals light lavender with dark purple base. Anthers in bottlebrush arrangement. Seeds with long silky hairs. Leaves pointed ovals with small, evenly spaced and sized teeth. Subshrub that looks like hibiscus .... *Hibiscus denudatus*

1' Flowers generally clustered along upright stem tips or 2–4 per leaf axil; if solitary, then smaller (less than 4 cm across) and not lavender. Colors various. Leaves generally broad, lobed, or unevenly toothed and wavy. Annuals, perennials, subshrubs, and shrubs.

  2. Subshrubs to shrubs.

    • Fruit cylindrical to spherical with 7–10 segments like orange segments with bristles or soft hairs and a short beak at the top. Petals orange .... *Abutilon palmeri*

    • Fruit disklike with 7–14 thin (2–5 mm) segments that split in half. Petals light purple .... *Malacothamnus fremontii*

    • Fruit with 8–12 segments, each with 2 wings at the top. Petals pink to red-lavender or yellow to pale orange (2 spp) .... *Horsfordia*

2' Annuals to perennials.

  3. Leaves fleshy, mainly at base of plant; sparse leaves along stem are deeply lobed into fingers. Plants with sparse or no hairs. Petals rose. Fruit segments 5–10 .... *Sidalcea neomexicana*

3' Leaves not fleshy, generally many along stems, lobes generally shallow or absent. Plants generally hairy. Fruit segments sometimes more than 10.

  4. Plants spreading along the ground.

    • Petals orange-pink to red, 3–6 mm .... *Abutilon parvulum*

    • Petals white or pale pinkish-purple, 4–6 mm .... *Eremalche exilis*

    • Petals creamy white to yellow, 10–15 mm .... *Malvella leprosa*

  4' Plants erect above the ground.

    • Petals magenta to rose, 20–30 mm, with a round spot (about 5 mm diameter) of purple at the base of each petal on the inside. Petals curving inward and overlapping, almost forming a sphere. Fruit segments 25–35 .... *Eremalche rotundifolia*

    • Petals white to pink, 4–5 mm. Fruits with about 11 segments .... *Malva parviflora*

    • Petals generally salmon orange to red-orange, but some almost white to lavender, from 7 mm to about 20 mm. Fruit segments 9–17 (6 spp) .... *Sphaeralcea*

Eremalche—This Greek genus name means "lonely mallow," as in being found in desert habitats where it has few companions. There are only three species, two of which are found in California's deserts. *Eremalche* are all annuals with only one seed per fruit segment.

Eremalche rotundifolia (**desert five-spot**) is an easy one to spot. The common name refers to the five distinctive purple or dark rose patches on the inside of each bright pink petal near the base—apparent only when looking down into the flower. Each petal is 2–3 cm long, nearly round, and deeply cupped, so that the whole flower is almost spherical and up to 4 or 5 cm in diameter. The Latin specific epithet, meaning "round" (*rotun-di*-) "leaved" (*folia*), is somewhat less helpful than the common name because many species in the family have somewhat round leaves. The distinctive flowers on an upright, frequently branched annual as much as 50 cm tall, with stems and leaves usually tinged purplish, is unmistakable. *Ere-malche rotundifolia* is found at low elevations throughout California's deserts.

Eremalche rotundifolia (desert five-spot)

Hibiscus (**rose-mallow, hibis-cus**)—*Hibiscus,* the Greek name that Dioscorides used for marsh-mallow, now applies to this largest of all the genera in the Mallow Family. Of the 200 (some say 300) species of *Hibiscus* in warm regions of the globe, only two are native to California, and only one of these grows in the desert. The genus *Hibiscus* differs from other desert mallow genera in not having true separate segments in the mature fruit. Instead, *Hibiscus* has a five-chambered fruit that spreads open forming a star, with the five parts remaining attached at the base.

Hibiscus denudatus (**pale face, rock hibiscus**) is a small subshrub. The Latin specific epithet means "stripped," "denuded," or having the leaves removed. In comparison to tropical species, this description may be apt, but for a desert plant it has plenty of leaves. The stems and leaves of *Hibiscus denudatus* are covered with densely intertwined, woolly, whitish hairs, making the leaves look as if they were cut from light-green felt with pinking shears. The flowers look just like a small version of the hibiscus flowers you see in gardens. Petals are about 2 cm long, spreading widely, and pale lavender, darkening to purple near the flower's center. The central bottlebrush of stamens surrounding the filament tube is red-orange, a beautiful contrast to the petals. As one common name suggests, this lovely hibiscus is often found in rocky canyons. Like its tropical relatives, it is frost sensitive, and thus is limited to fairly low elevations of the Sonoran Desert.

Hibiscus denudatus (pale face, rock hibiscus)

Sphaeralcea (**globemallow**)—The Greek word for "sphere" forms the root for this genus name, a reference to the spherical fruits. Although this general shape is shared by other genera in the family, the fruits of globemallow are perhaps a bit more spherical, as the common name itself underscores. It is characteristic of the genus for the spherical fruit to have 9–17 segments; each segment contains

Spheralcea ambigua var. *ambigua* (apricot mallow)

one to two seeds and breaks open only along the top portion of the segment, leaving the coarse bottom half intact.

    Sphaeralcea is a fairly large genus with about 50 species in warm parts of the Americas and in South Africa. Eight species, all natives, grow in California, six of which are found in the deserts. With the exceptions noted below, desert species have red-orange or salmon orange flowers with petals less than 2 cm long. There is one annual (*Sphaeralcea coulteri*), and the rest all are perennials, growing up to 1–2 m in height. Leaf shapes are highly variable in the genus, although most are some form of palmately veined. Leaves may be long and skinny, as in *Sphaeralcea angustifolia*; nearly triangular and three-lobed, as in *Sphaeralcea emoryi*; thick and smooth-edged, as in *Sphaeralcea orcuttii*; small and lobed all the way to the petiole, as in *Sphaeralcea rusbyi*; or broad and weakly three-lobed, as in *Sphaeralcea ambigua*.

Sphaeralcea ambigua (**apricot mallow**) has a species name that means "ambiguous," most likely a reference to the variability in leaf form and flower color. The most abundant and widespread of the varieties is *ambigua*, with apricot to orange-red flowers, often in spectacular clusters along erect or spreading stems. People pull off

Spheralcea ambigua var. *rosacea* (Parish mallow)

the road to take pictures of apricot mallow. Look closely and you will see hairs on the filament tube and yellow-purple anthers. The variety *rosacea*, sometimes called Parish mallow, has lavender to pink petals and purple-gray anthers. It is most commonly found in the Sonoran Desert. Finally, the variety *rugosa* has deep red-orange petals, similar to some shades of variety *ambigua*, but distinguishable by its wavy-edged, wrinkled-looking leaves and the lack of hairs on the filament tube. Its anthers are yellow.

Examples and Uses

Cotton (*Gossypium*) is the most commercially important member of Malvaceae. Cotton, cultivated as a fiber crop for thousands of years, seems to have been domesticated independently in Peru, Mexico, Africa, India, and China, with four different species being used. Cotton was planted extensively in southeastern North America soon after colonization, and the labor-intensive requirements of this crop became one of the justifications for slavery. When cotton fruits mature, the sections of ovary open to reveal long whitish hairs attached to each seed. Cotton fiber is valuable only with the seed removed, which is why Eli Whitney invented the cotton gin. The use of this machine to separate cotton fibers from their seeds saved thousands of hours of slave labor and boosted the early American cotton industry. Initially, seeds were used, if at all, mainly in livestock feed, but more recently oil from seeds has been used in margarine, prepared foods, and commercial cooking.

Okra (*Hibiscus esculentus*) is the only food crop plant in Malvaceae. In Latin, *esculentus* means "edible," which is a gross exaggeration. The part considered edible is an immature fruit with all the mucilaginous qualities for which the family is known. It is a rare chef who can make it palatable.

A more amazing culinary discovery was the use of mucilage from the roots of marsh mallow (*Alcea officinalis*), a pink-flowered marsh plant, mixed with eggs and sugar to make *pâte de Guimauve*, better known as marshmallow. As a shortcut to the French invention, gelatin is usually substituted for the mucilage and eggs, but the original is probably better.

Aesthetically important members of Malvaceae are such garden ornamentals as *Hibiscus* (numerous species, mainly tropical in origin), mallow (*Malva*), and hollyhock (*Althaea*).

NYCTAGINACEAE

FOUR O'CLOCK FAMILY

Abronia villosa (sand verbena), Sonoran Desert, Borrego Valley

Vibrant color characterizes the most common species of Nyctaginaceae. Sandy parts of the low desert after a good rain will spring to life with magenta flowers of sand verbena (*Abronia villosa*), growing from seed and rapidly covering large areas with mounds of color. Stay for the light of the setting sun and see *Abronia* glowing along with the surrounding mountains. This is one of the desert spectacles that people come over and over again to see.

The Greek prefix *nyct-* means "pertaining to the night" and describes the night-blooming habit of one of the largest genera of Nyctaginaceae. The common name four o'clock refers to the late-afternoon opening times for flowers of night-blooming species.

Nyctaginaceae Icon Features

- Perianth petal-like
- Perianth tube tightly enclosing a superior ovary
- One style, one ovary, and one ovule (seed) per flower

Distinctive Features of Nyctaginaceae

Nyctaginaceae species often have incomplete flowers, lacking petals, but with distinctive bracts just below each flower or flower group. This may be difficult to distinguish in cases where sepals look just like petals, and bracts like sepals. Using the term perianth for the colorful petal-like portion of flowers avoids the problem. Perianths in Nyctaginaceae are usually four- or five-lobed, or rarely six-lobed, even on the same plant. One rather eye-catching genus, *Allionia*, has a three-lobed perianth.

Ovary position is another point for confusion. Technically, Nyctaginaceae species have a superior ovary, positioned above the point of attachment of other flower parts. However, because the perianth (which looks like fused petals) forms an extremely tight tube that is shrunken around the ovary and often hardened, the ovary appears to be inferior.

Nyctaginaceae flowers either have radial symmetry or grow in well-defined sets that together have radial symmetry. Pistils have one ovary with one ovule, and one style with either a linear or a spherical stigma. There may be one to many stamens. Often the flowers are small and tightly clustered, with a bract holding them like a bouquet; larger flowers may appear singly or in small groups. Fruits are often ribbed or lobed or winged, sometimes in a bilaterally symmetrical pattern.

Forked branching and opposite positioning of leaves are good clues to Nyctaginaceae. The branching looks like numerous Y's strung together, because at each branching point both branches are equal. The leaves, though arranged opposite one another, frequently are not equal in size. Leaf shapes tend to be simple, with no lobes or teeth, and leaf blades are often somewhat fleshy. Many species are glandular hairy to the point of being unpleasantly sticky and covered with sand.

Similar Families

The closely related Pink Family (Caryophyllaceae) has opposite leaves and a few stem features and bracts similar to some Nyctaginaceae members, but pinks have either separate petals or no petals. If there are no petals, their sepals do not look like tubular corollas, as is the case with Nyctaginaceae.

Most other closely related families have desert representatives with no petals and no sepals posing as petals. In families with petals, they are separate and do not form a tube that resembles the petal-like calyx of Nyctaginaceae.

The Honeysuckle Family (Caprifoliaceae), which is not closely related, has tubular flowers and opposite leaves, but the California desert species are woody vines or shrubs, whereas none of the California desert Nyctaginaceae members are woody.

Sunflower Family (Asteraceae) members have several specialized features that superficially appear similar to those of Nyctaginaceae. The single inferior ovary of Asteraceae with a single ovule resembles the tightly encased superior ovary of Nyctaginaceae, also with a single ovule. The absent calyx and fused corolla of Asteraceae flowers appear similar to the colorful and fused calyx with absent petals in Nyctaginaceae. Both families have small flowers in tight clusters with leaf- or bractlike bases called an **involucre**. In other respects, however, the differences between the two families are substantial. Most desert species of Asteraceae that have opposite leaves, for example, also have radiate heads, or flowers of two kinds in a single head looking similar to a daisy. This arrangement does not resemble any members of Nyctaginaceae. In Asteraceae, the style has two branches, whereas there is no branching in Nyctaginaceae styles. Most Asteraceae species have fused anthers, whereas Nyctaginaceae anthers are separate.

Family Size and Distribution

There are 30 genera and about 300 species of Nyctaginaceae, found mainly in warmer regions of the Western Hemisphere. The largest genus, *Neea*, with 80 species, does not occur in California, but *Abronia*, *Mirabilis*, and *Boerhavia*, with roughly half as many species each, are represented in California, including the deserts.

In total, California has eight genera of Nyctaginaceae, all of which have native desert representatives. Many species seem to have an affinity for hot sandy places, but there are also several species restricted to desert mountains.

California Desert Genera and Species

California desert representatives of Nyctaginaceae are annuals, perennials, or subshrubs; they often are intricately branched and either upright or sprawling on the ground. Flowers are frequently some shade of pink, but white is common also, especially in night-blooming genera. Sometimes individual species may have either pink or white flowers.

SIMPLIFIED KEY TO DESERT GENERA OF NYCTAGINACEAE

1. Perianth 3-lobed, with 3 bilaterally symmetrical flowers blooming as 1 dish-shaped radially symmetrical flower with 3 magenta petals. Annual, trailing on ground .... *Allionia incarnata*
1' Perianth with more than 3 lobes or no apparent lobes, but plenty of tube. Radial symmetry.
  2. Perianth tiny (2 mm or less in length) (5 spp) .... *Boerhavia*
  2' Perianth small to large (5 mm or more in length).
    3. Perianth 30 mm or longer and flowers few (1–6) per cluster.
     • Perianth 70 mm or more, narrow, white near the tip .... *Acleisanthes longiflora*
     • Perianth 30–40 mm, narrow, white near the tip .... *Selinocarpus nevadensis*
     • Perianth 40–60 mm, widely spreading, magenta .... *Mirabilis multiflora*
    3' Perianth less than 30 mm or if slightly longer then flowers in clusters of more than 10.
      4. Perennials growing erect and lanky. Leaves mainly below. Stems little branched and bare looking.
       • Stems with a dark sticky ring, like a collar, between branching points and leaf pairs. Perianth pink, 8 mm .... *Anulocaulis annulatus*
       • Stems without collars. Perianth red, 15–20 mm .... *Mirabilis coccinea*
      4' Annuals, perennials, and subshrubs, often with stems partially supported on the ground, densely branched, and leafy throughout. No collarlike rings on stems. No red perianths.
        5. Perianth spreading widely from a short tube, like a funnel or saucer. Stamens extending in front of perianth and clearly visible (7 spp) .... *Mirabilis*
        5' Perianth mostly tube with small lobes spreading just at the tip. Stamens inside tube and not visible.
         • Fruits with 3–5 broad, membranous wings, almost transparent. Wings joined above and

below seed chamber and continuous all the way around .... *Tripterocalyx micranthus*
• Fruits with 0–5 wings, sometimes spongy or hollow; some thin but not transparent. Wings not joined above the seed chamber (5 spp) .... *Abronia*

Abronia (**sand verbena**)—This genus is named "graceful" in Greek for the way stems dance across the desert sand, growing in a single season to almost a meter in length and producing a profusion of flowers. Stems are usually reddish and glandular, and leaves somewhat fleshy—features not normally considered "graceful," but in this genus the overall effect can be stunning. Numerous trumpet-shaped flowers appear in clusters that stand above the bulk of stems and leaves, coloring the desert floor following sufficient rains. Perianths are 6–35 mm in length, and lobes spread to a width that is roughly half the length of the tube. Fruits often have wings, the shape and number of which help to distinguish species. The "verbena" of the common name is a misnomer, implying an entirely different family (Verbenaceae), but the "sand" qualifier places this flower in its favorite habitat, at least for desert species: the open sand. One species, *Abronia nana*, is a small tufted perennial favoring desert mountains.

Abronia villosa (**sand verbena**)

Abronia villosa (**sand verbena**) is a glandular-hairy annual, *villosa* being Latin for "hairy." It grows to about 1 m, with its main stems hugging the ground and the flowering branches forming mounds. It is amazing that an annual can spring to life, grow to such a size, and flower in such profusion so quickly following rain, after which the seeds can wait decades in the desert sands for the next good rain. *Abronia villosa* flowers are pink to magenta in clusters of 15–35. Fruits usually have three to five thin wings, but there may be none. It grows in both deserts in loose sand, at elevations below 1,600 m.

Allionia (**windmills, trailing four o'clock**)—This American genus, having just two species, was named for the Italian botanist Carlo Allioni (1725–1804). Its characteristic spread across the ground provides the adjective "trailing" for one of the common names. The bilaterally symmetrical flowers grow in clusters of three and bloom together, giving the illusion of a single flower with radial symmetry and three petals, each of which has three lobes.

Allionia incarnata (**windmills, trailing four o'clock**) is the only species of *Allionia* found in California. Its perianths are usually bright magenta, although

Allionia incarnata (**windmills**)

they may be lighter in color, as suggested by the specific epithet *incarnata*, meaning "of the flesh"—or here, flesh-colored. The stems and sparse leaves are covered with tiny grains of sand that stick to numerous hairy glands. Fruits are cupped, with the incurved sides usually having a few teeth and the recessed center having two rows of sticky glands. Windmills likes sandy places below 1,500 m.

Anulocaulis (**ringstem**)—This is another small genus with just five species, living mostly in Mexico; only one ranges into California. It takes its name, *anulo-* (ring) plus *caulis* (stem), from the sticky dark rings that form around the stem roughly halfway between each point of branching or leaf attachment. This genus consists of perennials with few branches and leaves that are concentrated at the base, giving the 1.5-m-long stems a lanky look. The stamens, usually three to five in number, extend well outside the perianth.

Anulocaulis annulatus (**sticky ringstem**) is twice named for its rings because *annulatus* means "marked with a ring." The common name refers to the glandular nature of the rings with their sticky discharge, which may discourage insects from climbing the stem to eat tender flower buds. Leaves of sticky ringstem are largest and most numerous near the base of the plant; they are up to 10 cm long and sometimes almost round or heart-shaped, with stiff hairs having dark glandular bases. The small (8 mm) pink flowers are tubular, with hairs around the lower portion of the tube; they grow in tight clusters at the ends of slender branches. Sticky ringstem is most common in Death Valley, where it frequents road cuts and rocky slopes or canyons below 1,200 m.

Anulocaulis annulata (sticky ringstem)

Boerhavia (**spiderling**)—This genus of about 30 species is named for the Dutch botanist and professor of medicine Hermann Boerhaave (1669–1738). "Spiderling" describes the lanky, jointed appearance of the hairy stems with sparse leaves. There is often a sticky area around the stems between the branching joints, but it is not dark and distinct as in *Anulocaulis*. Flowers are tiny (2 mm or less) in all five California desert species, and they grow in small clusters with one to three bracts below the group. There are one to five stamens per flower, and a spherical stigma, but the flowers close in the afternoon, so get there early! Flower color, stem hairs and glands, and fruit shape and texture help to distinguish species.

Boerhavia wrightii (**spiderling**) is the most widespread of the species, growing in dry sandy places below 1,400 m. It is a glandular-hairy annual growing to 70 cm in a

Boerhavia wrightii (spiderling)

Mirabilis bigelovii (wishbone bush)

sparsely branched, open arrangement like spidery legs. The stems, leaf edges, and bracts below flowers are red or red-tinged, and flowers are pink or red. Fruits are a little longer than they are wide and have four lobes, or fat ridges. The species is named for Charles Wright (1811–1885), who collected plants from Texas to California and the Pacific Northwest, documenting many new species.

Mirabilis (**four o'clock**)—Sixty or more species of *Mirabilis* are distributed throughout the Americas and the Himalayas, with 12 species in California (several nonnative), eight of which occur in the deserts. *Mirabilis* species are perennials or subshrubs that are heavily branched in the Y arrangement. One to many flowers, blooming sequentially, grow as a bouquet held by a cup-shaped involucre, which looks like fused sepals serving the whole group of flowers rather than belonging just to one. Flowers open around four o'clock in the afternoon and close in the morning. Fruits are round to club-shaped and may have lobes, but not wings. The genus is named "wonderful" (*mirabilis* in Latin), so go take a look.

Mirabilis bigelovii (**wishbone bush**) is a common species with two intergrading varieties. The specific epithet honors Professor Jacob M. Bigelow (1786–1879) of Harvard Medical School, who authored *American Medical Botany.* Wishbone bush has glandular-hairy green stems forming a rounded bushy-looking plant up to 80 cm high. The Y-type branching, with two equally sized branches forming at each junction, looks somewhat like "wishbones" sprouting throughout the plant—hence the common name. Flowers, one per involucre, are generally numerous, about 1 cm across, white (or sometimes pink), and widely funnel-shaped. *Mirabilis bigelovii* grows in rocky spots throughout the deserts at elevations below about 2,300 m.

Mirabilis coccinea (**scarlet mirabilis**) has a specific epithet describing its "scarlet" (*coccinea*) flowers—unusual for the genus. Flowers are narrow and up to 2 cm long. Stems are thin, forked, and almost leafless, rising to

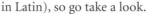

Mirabilis coccinea (scarlet mirabilis)

about 50 cm. Look for it on rocky slopes or washes of the eastern Mojave Desert from 1,300 to 1,800 m.

Mirabilis multiflora (**giant four o'clock**) is "many" (*multi-*) "flowered" (*flora*), and the blossoms, six per involucre and magenta, are "giant"—up to 6 cm long. Stems are up to 80 cm in length and usually rest on the ground. Leaves are large too—up to 12 cm long—oval to round, and a bit fleshy. They may loose their glandular-hairy coating with age. Giant four o'clock grows in rocky or sandy spots throughout the deserts at elevations below 2,500 m.

Examples and Uses

Bougainvillea, the electric-colored vine adorning trellises and balconies in hot climates, is perhaps the most familiar member of Nyctaginaceae. The colorful part of *Bougainvillea* is the three bracts that surround the three tiny flowers, forming a unit that looks, to the uninitiated, like a flower. *Bougainvillea* is native to South America and has not naturalized in California. Other ornamentals in the family include four o'clock (*Mirabilis*) and sand verbena (*Abronia*), many of which require mild climates.

A few species are edible or medicinal but are not of commercial or significant local importance.

Mirabilis multiflora (giant four o'clock)

Oenothera deltoides (dune evening primrose), Mojave Desert east of Twentynine Palms

ONAGRACEAE

EVENING PRIMROSE FAMILY

Many members of Onagraceae are familiar inhabitants of classic sandy desert habitats. Sand dunes would not be complete without a spring complement of dune evening primrose, annuals with large white petals and the ethereal look of something that cannot last. Their season is short, but year after year they grow from seed and remind us that a little water goes a long way in the desert.

Onagros is the Greek word for a wild ass of dark yellowish coloring, native to areas near the Caspian Sea. What this animal has to do with a family of delicate flowers is unclear. The name evening primrose makes more sense, referring to the habit of many flowers in this family to open in the evening for pollination by moths. Primrose as part of the common name, however, is misleading: the true Primrose Family, Primulaceae, is only distantly related to Onagraceae.

Onagraceae Icon Features

- Four petals in radial symmetry (some exceptions), often two-lobed
- Eight stamens (rarely four or 10) with large anthers
- Central hypanthium elongated (often) above an inferior ovary

Distinctive Features of Onagraceae

An inferior ovary is an invariable feature of Onagraceae. In addition, a **hypanthium**, or floral cup, frequently (but not always) forms a nectar-bearing tube between the ovary and the attachment of the sepals, petals, and stamens. The style reaches through this tube from its attachment to the ovary to extend the stigma into a visible position with the petals.

Most species of Onagraceae have four petals (rarely five) that are often large and showy. Petals are free, but all of them are attached to the tubular hypanthium, so when they fall they fall together. Flower symmetry is usually radial, but there are several exceptions, including one desert genus, *Gaura*, and some almost-bilateral members of *Epilobium*. There are four sepals (rarely five) and either four or eight stamens (rarely 10), often with large anthers. The stigma frequently is distinctive, either with four long lobes or with an enlarged club or ball shape big enough to hang of its own weight to below the center of the flower. The shape of the stigma and length of the hypanthium may be used in determining genus and species.

The inferior ovary is divided into four (sometimes two) long chambers, generally with many seeds. In some genera, seeds have long or fluffy hairs attached to improve their airworthiness during dispersal. Even though most Onagraceae

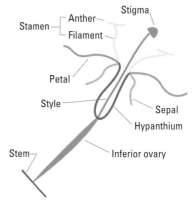

Evening Primrose Family (Onagraceae) flowers generally have a long tubular hypanthium attached above the inferior ovary.

species are designed to attract pollinators, during either the day or the night, many are self-pollinating with flowers that never fully open, day or night. This feature may be useful in identification, but can be frustrating to photographers.

Many species of Onagraceae are annuals or perennials, but the family also includes woody members that are shrubs or even trees. Even annual members of the family sometimes have stems that peel near the base, as if a thin layer of bark were splitting and curling back to fall off the stem. This feature is helpful in identifying species. Leaves are highly variable but tend to be simple shapes, with teeth or lobes that may be shallow or deep. Leaf arrangement is also variable and includes alternate, opposite, and whorled attachments, without stipules or with stipules that fall off early in development of the leaves.

Similar Families

Mustard Family (Brassicaceae) and some Poppy Family (Papaveraceae) members have four free petals in radial symmetry, but their ovaries are superior, conspicuously growing above the point of petal attachment. The Loosestrife Family (Lythraceae) is closely related to Onagraceae and sometimes has four petals in radial symmetry. Like mustards and poppies, loosestrife has a superior ovary, but a subtle technicality could confuse you: the hypanthium in Lythraceae, which forms a base for the petals, grows up around the superior ovary so that the petals appear above the ovary. But petal attachment is the key, and when a hypanthium is present, it should be considered as if it were a lower extension of the petals. In the case of Lythraceae, the ovary is above (superior to) the attachment of the hypanthium. In Onagraceae, the ovary is beneath (inferior to) the attachment of the hypanthium.

Another difference to look for is the number of stamens. Mustards generally have six stamens, four of them long and two short (so check carefully for these smaller ones), rather than the four or eight stamens of evening primroses. The only desert poppies with four petals also have 12 or more stamens—a feature you will not see in any evening primroses. Lythraceae species, however, have four, six, or occasionally more stamens.

Distribution

Depending on whom you ask, Onagraceae comprises 15 to 18 genera, distributed throughout the world in approximately 650 species. Western North America has a heavy representation of species, especially in the two largest genera, *Epilobium* (willow herb) and *Oenothera* (evening primrose). The third largest genus, *Fuchsia*, is largely a tropical shrub but is widely cultivated in mild climates.

California hosts eight genera of Onagraceae, six of which are represented in the deserts. Although the genera *Epilobium* and *Clarkia* are abundant throughout most of California, only three species of the former, and none of the latter, grow in the deserts. By far the greatest numbers of desert species are in the genera *Camissonia* (sun cup) and *Oenothera*. With rare exceptions, desert species of Onagraceae are natives.

Both *Camissonia* and *Oenothera* consist primarily of species favoring sandy habitats, including sand dunes and other inhospitable-looking locations. Another genus, *Ludwigia*, with only two desert representatives, requires wet habitats, and still others, such as *Gaura* and *Epilobium*, grow mainly in the mountains.

California Desert Genera and Species

All desert species of Onagraceae are annuals or perennials with one exception: California fuchsia (*Epilobium canum*) is a subshrub in California's desert mountains. Most desert species have flowers that are either white, especially the night-blooming ones, or yellow. If there are just a few flowers per stem, they tend to be large, measuring 5 or 6 cm across. Smaller flowers of roughly 1 cm often grow in elongated clusters. As usual, there are exceptions, including some very tiny (less than 2 mm) single flowers in the genus *Gayophytum* and a few small-flowered species of *Camissonia* and *Epilobium*.

SIMPLIFIED KEY TO DESERT GENERA OF ONAGRACEAE

One of the most useful features for determining genus in Onagraceae members is the shape of the stigma. Although shape is not entirely consistent throughout all genera, it is a main feature of most keys.

Hemisphere
Camissonia

Sphere
Camissonia
Ludwigia
Gayophytum

Club
Epilobium
Ludwigia

Long lobes
Epilobium
Oenothera
Gaura

Different genera of the Evening Primrose Family (Onagraceae) have stigmas with characteristic shapes.

1. Stigmas deeply 4-lobed, lobes generally linear.
  2. Flowers with distinct bilateral symmetry (2 spp) .... *Gaura*
  2' Flowers with radial symmetry, rarely slightly bilateral.
    • Petals white or yellow and 5 mm or more, some fading red to purple (8 spp) .... *Oenothera*
    • Petals pink, magenta, or red; if white, then 1–3 mm (3 spp) .... *Epilobium*
1' Stigmas spherical, hemispherical, or club-shaped.
  3. Stamens 4 or 10 (2 spp) .... *Ludwigia*
  3' Stamens 8, sometimes 4 longer than the others.
    4. Extensively branched annual with threadlike stems. Petals white, less than 2 mm. Ovary 2-chambered. Growing above 1,800 m in the Panamint Mts. .... *Gayophytum decipiens*
    4' Stems generally not threadlike, and extensively branched. Petals white, yellow, or rarely pink or lavender. Ovary 4-chambered. Most not in the Panamint Mts.
      • Stigma generally club-shaped. Streambanks and moist habitats. Petals white or pink (2 spp) .... *Epilobium*
      • Stigma spherical or hemispherical. Dry habitats. Petals white, yellow, rarely lavender (some fading reddish) (19 spp) .... *Camissonia*

Camissonia (**sun cup**)—The German botanist Adelbert von Chamisso (1781–1838) is honored by this genus name. When you see bright yellow *Camissonia* flowers in the sunlight, however, you will remember the common name, sun cup. There are 62 species of *Camissonia*, primarily in western

North America, with one in South America. Forty-three species, all natives, are found in California, 19 of which are represented in the deserts. Almost all desert species of *Camissonia* are annuals, so their flowering is entirely dependent on good rainfall. When growing in clusters, these flowers may dominate a desert landscape after unusually heavy spring rains. *Camissonia* flowers usually are open during the day, and they have distinctive spherical or hemispherical stigmas, extending beyond the anthers or hanging below the center of the flower. This trait, in combination with four white or yellow petals, is an excellent indicator of the genus. Desert *Camissonia* typically are under about one-half meter in height, and they have basal clusters of leaves or alternate arrangements of stem leaves, or both. Leaves are frequently long and somewhat narrow in general outline, but are often lobed or even pinnately divided into numerous leaflets along the central axis of the blade. Even on a single plant, there may be great variation in leaf lobe shapes and sizes.

Camissonia boothii (woody bottle-washer)

Camissonia boothii (**woody bottle-washer**), widespread in the deserts, is readily identified by its curled spike of white and reddish flowers, with numerous simple leaves on the stem immediately below. Although the flowers officially open at dusk, the appearance of the flowering stem is quite recognizable during the day, and some flowers are usually open. Older flowers, stems, and even buds tend to be reddish (or pink), while the freshly opened flowers are usually white. There may be basal leaves as well as leaves along stems, and there are sometimes two or three flower spikes from the same set of basal leaves. The common name, woody bottle-washer, refers to the hard dried stems of the late season that still retain numerous spirally arranged, long, skinny, split-open fruits, as if they were the bristles on a bottle brush. The specific epithet honors the otherwise obscure M. B. Booth. Peter Raven, presently the director of the Missouri Botanical Garden, has described six sub-species of *Camissonia boothii* in California alone. These intergrade widely, and include as sub-species several forms formerly recognized as species.

Camissonia brevipes (**yellow cups**) is one of the more spectacular *Camissonia* members in a good year. Even in Death Valley it has been known to provide a profusion of flowers across acres of stony alluvial slopes that have been almost bare for decades. The four yellow petals are usually about 1 cm long, with tiny red dots

Camissonia brevipes (yellow cups), Death Valley National Park, near Hole in the Wall

at their bases, and together look like yellow cups—hence the common name. The stems and main leaf veins tend to be quite reddish. Well-watered plants grow to a half meter or more, with most of the leaves at the base, so that nodding flower clusters at the stem tips are highly visible. *Brevipes* means "short-footed" in Latin and describes the short (2–20 mm) stalk that attaches the long (2–10 cm) fruit to the stem. There are three subspecies in California, and some of these hybridize with other species to make identification particularly frustrating.

Camissonia campestris (Mojave sun cup)

Camissonia campestris (**Mojave sun cup**) is a small (5–25 cm) plant with very thin, delicate stems and leaves. It would hardly be noticed without its out-of-proportion yellow flowers, up to 3 cm across and seeming too big to be supported by the slender, bending stems. Each petal generally has two red dots at its base, and flowers fade to a reddish color. The Latin *campestris,* meaning "of the fields," describes the open sandy flats or grasslands where this species is found. Its habitat includes just the western part of the Mojave Desert, providing the necessary justification for including "Mojave" in the common name. As usual, there are intergrading subspecies as well as hybrids with other species.

Camissonia claviformis (**brown-eyed primrose**) is an annual with numerous different forms, blooming profusely

Camissonia claviformis ssp. *claviformis* (brown-eyed primrose)

Camissonia claviformis ssp. *piersonii* (Pierson brown-eyed primrose)

following good early-season rains. Although flower color may be either white or yellow, most forms have brown or dark purple spots ("eyes") at the base of the petals, large enough to make a dark center to the flower. Leaves are usually basal and deeply lobed down to the central axis, but with a much larger lobe at the tip of the leaf. *Claviformis,* meaning "shaped like a club," refers to the fruit, which is 1–3 cm long and slightly wider near the tip than at the base, like a baseball bat, although it may be either straight or curved. There are 11 subspecies, with eight in California, generally intergrading and sometimes hybridizing with other species. *Camissonia claviformis* can be found under 2,000 m in sandy or rocky places throughout the deserts.

Camissonia refracta

Camissonia refracta is one of the few white-flowered desert *Camissonia* species. Flowers open at dusk and are only about 1 cm across (each petal 4–7mm). It is a small annual (about 20 cm) with fairly small (less than 6 cm), thin, unlobed leaves growing along the stem, and no basal leaves. *Camissonia refracta* (the specific epithet meaning "broken open") grows mainly on sandy flats or slopes below 1,300 m in both deserts.

Epilobium (**willow herb**)—The name of this genus characterizes the family feature of the inferior ovary: *epi-* means "on top of," indicating that the other flower parts are on top of the *lobium,* the "pod" or "fruit." The common name refers to the herbaceous habit of most members of the genus, and to the willow-shaped leaves that are typical. Although this is a very large genus, with 171 species worldwide and 30 in California, only four occur in our deserts—and all are very different from one another. *Epilobium* species have four sepals and four petals like the rest of the family, and eight stamens. The stigma is usually club-shaped, but two of our desert species have four-lobed stigmas, and a third desert species sometimes has a four-lobed stigma. The most distinctive, although not completely reliable, feature of the genus is white hair or fluff attached to each seed. Petals are often pink, magenta, or rose-colored, but white, red, and other variations are available. Petals are often notched, or two-lobed, although this feature is not exclusive to *Epilobium.*

Epilobium angustifolium **ssp.** *circumvagum* (**fire-weed**) is found at northern latitudes throughout the world and makes a California desert appearance only in some northeastern desert mountains of the state. Otherwise, it is not considered a desert species, but rather is typical of mountainsides, especially following disturbance by fire (the common name refers to this habitat preference) or road-grader. The specific epithet is a compound Latin name meaning "narrow" (*angusti-*) "leaved" (*folium*)—not to be confused with *august,* which means "wise", "dignified", or "splendid."

Epilobium angustifolium ssp. *circumvagum* (fireweed)

Fireweed is a perennial with upright stems often over a meter tall, topped by a spike of bright magenta flowers. Patches of fireweed are noticeable from a distance. Leaves, alternately arranged up the stem to where the flowers start, are 15–20 cm long but only 1–2 cm wide (i.e., *angust*). There are dozens of flowers in a spike, with the buds pendant at the top and older flowers forming long fruits at the bottom of the spike. Each flower is about 2 cm across, and the stigma has four lobes. Check maturing fruits for seeds with white hairs attached.

Gaura—The 21 species of *Gaura* are concentrated in Texas, but four have made it into California. Only one, *Gaura coccinea,* is native in our state, and it is found in our desert mountains. *Gaura sinuate* is an invasive weed that has become established in the Mojave Desert. *Gaura* means "proud" or "majestic" and describes the eye-catching quality of many species in the genus.

Gaura coccinea (**scarlet gaura, wild honeysuckle, linda tarde**) has bilaterally symmetrical flowers that turn red with age. Petals of fresh flowers are white, but red anthers, fruits, and older flowers on the same stems give the overall impression of "scarlet," or *coccinea*. Plants are perennial, with a woody caudex at ground level. Scarlet gaura grows at 900–1,600 m, mainly in the eastern desert mountains.

Oenothera (**evening primrose**)—This is the genus that carries the common name for the family, and that typically has the largest flowers. *Oenothera*, Greek for "wine-scented," refers to the pleasant fragrance of some members of the genus. There are 119 species, all native to the Western Hemisphere; 17 species occur in California, though only 11 of these are native to the state. California deserts have eight of the state's total species, only one of which is nonnative.

Desert *Oenothera* have white or yellow petals that are notched or toothed at the tips into short, fat lobes, like the top of a valentine. Most flowers close during the day, looking similar to buds. At dusk, sepals reflex rapidly, almost popping back, and petals spread within a minute or so. Usually, moths are already waiting nearby, their long proboscises (drinking straws) ready to access nectar deep in hypanthium tubes. Early morning is the best time to look at *Oenothera*, before flowers close but while light is good.

Gaura coccinea (scarlet gaura)

Oenothera have eight stamens, each with a long anther attached in the middle, forming a T-shape with the filament. The stigma also has four long lobes, so the inside of the flower has numerous leggy-looking parts. In fact, there is a leggy spider, colored the same as the flower parts, that virtually disappears among anthers and stigma lobes of *Oenothera deltoides*, waiting for insects attracted by the fragrance.

Desert *Oenothera* may be annuals, biennials, or perennials. Leaves may be basal with essentially no stem, the flowers simply growing out through the dense clump of leaves. Alternatively, there may be many leaves spaced out along obvious curving, sprawling stems up to nearly 1 m in length. Leaf shape is similar to *Camissonia*, and includes numerous variations on lobes.

Oenothera californica (**California primrose**) is a white perennial that blooms during the evening and night hours. In this species, buds are nodding, or down-turned due to a bend in the hypanthium. After starting out as a single cluster of leaves with a few flowers, *Oenothera californica* grows a peeling stem and branches, low to the ground, so that a single plant may

Oenothera californica ssp. *avita* (California primrose)

cover almost a meter of ground and have numerous flowers. Leaves grow both at the base and along stems; up to 6 cm long, they are narrow ovals, usually without lobes, and generally with a gray-green cast. Fruits are 2–8 cm long and only 2–3 mm wide. *Oenothera californica* grows in sandy or gravelly places under about 2,500 m. There are three subspecies, including the endangered *eurekensis*, the Eureka Dunes primrose, which is found only in that general area near Death Valley. It is distinguished by having new clusters of leaves forming at the stem tips, and by having fleshy roots—though it would be a federal crime to dig them up to look at the roots!

Oenothera deltoides (dune evening primrose)

Oenothera deltoides (dune evening primrose, devil's lantern, lion-in-a-cage, basket evening primrose) has three common names that describe the tendency of older stems to curve upward and inward in a basketlike formation. Otherwise it is much like *Oenothera californica*, starting out with a single cluster of basal leaves, growing a peeling stem, and having nodding buds and white flowers that open in the late afternoon. However, *Oenothera deltoides* has wider and larger leaves than its cousin, not necessarily *deltoid*, or triangular, but often somewhat diamond-shaped and lobed. Also, *Oenothera deltoides* is highly fragrant, whereas *Oenothera californica* has only a faint aroma. In many cases, habitat alone will distinguish the two species, with *Oenothera deltoides* growing in very sandy areas at elevations under 1,800 m. There are five subspecies, only two of which occur in California's deserts.

Oenothera primiveris ssp. *primiveris* (yellow evening primrose)

Oenothera primiveris (yellow evening primrose) is a low annual with leaves in a single basal cluster and yellow flowers that bloom in very early (*primi-* or "first") "spring" (*veris*). Leaves are usually 10–15 cm long and may be any shape from entire to pinnately lobed. There are two subspecies: *bufonis*, with grayish green leaves and petals 3–4 cm long; and the less common subspecies *primiveris*, with green leaves and petals only 0.5–2.5 cm in length. Either subspecies may be found in sandy areas below 1,400 m.

Examples and Uses

Onagraceae has many ornamental members. *Fuchsia* species and cultivars have distinctively shaped pendulant flowers and are grown in shaded parts of the garden as shrubs or in hanging baskets. They are frost sensitive and have recently been decimated by a virus that limits their availability in nurseries. Evening primroses are tough and reliable as showy garden plants, often doing well with minimal watering. *Clarkia* members are colorful annuals, including farewell-to-spring, red-ribbons, and mountain garland. Many of these have been bred to have larger, more colorful flowers than their wild relatives.

No common foods are derived from Onagraceae, and there are no commercial uses of which we are aware, other than as ornamentals.

The night-blooming feature of several members of the family makes them attractive to moths. For example, the white-lined sphinx moth (*Hyles lineata*) depends heavily on Onagraceae, both for night feeding on nectar by adult moths and for all-day eating by the caterpillar portion of the life cycle.

Eschscholzia californica (California poppy), Antelope Valley California Poppy Preserve

PAPAVERACEAE

POPPY FAMILY

The most prominent member of Papaveraceae is the California poppy, our state flower, notable for its brilliant orange color and sometimes blanketing entire hillsides following a good rainy season. However, California poppies thrive on the moist fringes of our deserts, and it is their lighter-colored cousins that are the better adapted as desert dwellers. Many other poppies are ornamentals in our gardens, valued for their brilliant early bloom. Although oriental poppies have a bad reputation as the source of opium and heroin, the large and colorful flowers must be a fine spectacle in the fields of Afghanistan and other lands where their cultivation is an irresistible source of revenue.

Papaver is the Latin word both for poppy and for the genus that lends its name to the family. Although the common name is probably just a distortion of the Latin, there is another way to think of it: poppies usually have two sepals that "pop" off just as the flower opens.

Papaveraceae Icon Features

- Large flower buds with two (sometimes three) sepals that split apart and fall away as the flowers open
- Superior ovary, short or nonexistent style, and enlarged stigma
- Stamens usually numerous

Distinctive Features of Papaveraceae

Popping pairs of sepals are a good indication of Papaveraceae. Usually there are just two sepals that split apart abruptly and fall away to let the opening petals steal the show, though some genera have three or four sepals, and sometimes the sepal pairs remain united as they fall. But still, Papaveraceae members have a certain look. The sepals tend to be fairly large, enclosing four or six separate, nearly mature petals, brightly colored or white, that spread rapidly when the sepal casing releases them. Symmetry is radial in most genera, including all California desert species, but some nondesert genera have bilateral symmetry.

Papaveraceae species generally have one superior ovary with one chamber. Sometimes it is long and thin, sometimes short and fat, and occasionally it is segmented. Often there is no style at all, or a very short one, so that the stigma (with or without lobes) sits right on top of the ovary. Most desert genera have many small seeds, but the family as a whole has no rules about seed numbers. There are usually numerous stamens, but a few genera have only four, and one desert genus, *Canbya*, has six.

Sap is often milky or yellow- to orange-colored in Papaveraceae members. Leaves are often dissected or deeply lobed, and occasionally they are stiff and prickly. Plants in general may be annuals, perennials, shrubs, or small trees, but our deserts have only annuals and perennials.

Similar Families

The largest desert family having flowers with radial symmetry and four separate petals is the Mustard Family (Brassicaceae), but its petals are usually clawed, or bent like an "L" in the middle, which is not the case in any Poppy Family desert members. Typically, Mustard Family flowers are small and arranged in large clusters, as opposed to the larger, frequently solitary Papaveraceae flowers. Finally, mustards never have numerous stamens, as is typical of poppies.

Evening primroses (Onagraceae), with four large white or yellow petals in radial symmetry, are superficially similar to some Papaveraceae members, but there are several reliable differences. Evening primroses always have an inferior ovary, as opposed to the superior ovary of the Poppy Family. Evening primroses have either four or eight stamens, as opposed to many (or six), and evening primroses generally have four sepals that do not fall away as the flower opens, as opposed to two (or three) sepals that pop away as the flower opens.

Some Purslane Family (Portulacaceae) members have two sepals, but they generally do not fall away as the flowers open. Purslanes tend to be fleshy, and often the flowers are tiny, neither of which characters is common in desert poppies.

Family Size and Distribution

Papaveraceae, which embraces 40 genera and 400 species, is most abundant in subtropical and temperate zones of the Northern Hemisphere, though a few species are found in Africa and Australia as well. The largest genera are *Corydalis* and *Papaver*, neither of which is represented in California's deserts, although they are frequently cultivated.

California has 14 genera of Papaveraceae, most of which are small genera with few representatives in the state and two of which comprise entirely nonnative species. Our deserts are home to just five genera with nine species of poppies, all natives.

In California's deserts, Papaveraceae species are probably most abundant in the western Mojave Desert, where rains are somewhat reliable. The Antelope Valley California Poppy Reserve, near Lancaster, is a desert grassland that becomes a sea of orange when California poppies (*Eschscholzia californica*) are in full bloom. Cream cups (*Platystemon californicus*) also are abundant there. Although several other species of Papaveraceae are more widespread in our deserts, they seldom compete in abundance with a good bloom year in the Antelope Valley. A few species of poppies are particularly successful within a few years of a burn.

California Desert Genera and Species

Desert species of Papaveraceae in California are either annuals or perennials, usually with fairly large (2–8 cm) flowers appearing either singly or in small groups at the ends of single stems or short branches. Petals are white, cream, yellow, or orange. Some plants have no hairs, while others have long, distinctive hairs or even prickles.

SIMPLIFIED KEY TO DESERT GENERA OF PAPAVERACEAE

All California desert genera of Papaveraceae have flowers with radial symmetry and either four or six petals. Sometimes individual plants within a single population of a species will have either four or six petals per flower, so it is important to look at several representatives. Genera are easily identified by simple characteristics of leaf shape, spines, and hairs as well as numbers of stamens. These features are more reliable than petal numbers.

1. Leaves and stems totally without hairs but sometimes having a whitish powdery coating. Leaves deeply dissected into almost linear segments. Petals 4 or sometimes 6 (4 spp) .... *Eschscholzia*

1' Leaves and stems generally hairy or spiny. Petals generally 6, some 4. If hairs are sparse, petals must be 6. Leaves entire, toothed, or lobed, but not dissected into linear segments.

  2. Leaves toothed and lobed with sharp spines that hurt when you touch them. Leaves positioned all along the stem, but usually larger near the bottom (2 spp) .... *Argemone*

  2' Leaves hairy and sometimes toothed but without sharp spines or deep lobes. Leaves positioned only at or near the base of the plant.

    3. Leaves wedge-shaped, distinctly wider and toothed at the end away from the stem. Petals either 4 or 6, large (2.5–4 cm), and white. Stamens many .... *Arctomecon merriamii*

    3' Leaves without teeth or lobes and narrowly oblong to linear. Petals 6, small to medium (3–16 mm), and white or cream to yellow. Stamens 6–9 or more than 12.

      • Leaves almost without hairs, somewhat fleshy. Petals 3–4 mm, white. Stamens 6–9 .... *Canbya candida*

      • Leaves with long dense hairs, not fleshy. Stamens more than 12. Petals 20–30 mm, cream to yellow .... *Platystemon californicus*

Argemone (prickly poppy)—Prickly is the right name for this genus. Members are well armed with sharp spines, especially on the teeth of leaves that grow all the way up the stem. The Greek word *argema* is a disease of the eye, and the scientific name *Argemone* refers to an herb supposedly used as a cure for ocular cataracts. There is no claim that this genus has any such properties. In fact, the sap of these plants is acrid and a milky yellow or orange—not the sort of thing to put in your eye! There are about 30 species in North and South America and Hawaii. Only two species occur in California, and these are both desert residents.

Argemone munita (chicalote) is the larger of the two very similar Cali-

Argemone munita (chicalote)

Eschscholzia glyptosperma (desert gold-poppy)

fornia species of *Argemone*. It can grow to 1.5 m but is usually under 1 m. Its six white petals are 2.5–4 cm long and at least as wide, forming a complete white circle around a huge cluster of up to 250 golden-orange stamens. The sap of *Argemone munita* is yellow and toxic, but it is difficult to imagine anything trying to eat this spiny plant. The Latin word *munita* means "provided with," in the sense of being "fortified" or "protected"—an apt description. *Argemone munita* grows on open slopes of desert mountains at elevations up to 3,000 m. Some of the most spectacular showings are in areas recently burned.

Eschscholzia—This unpronounceable name honors the Russian naturalist and surgeon Dr. Johann Friedrich Gustav von Eschscholtz, who traveled with explorers looking for the Northwest Passage. They stopped for provisions in the San Francisco Bay area in 1815, allowing time for plant collections that included the California poppy (*Eschscholzia californica*). There are 12 species of *Eschscholzia*, all in western North America. Ten of them occur in California, and four of these are represented in our deserts.

All *Eschscholzia* have leaves that are dissected into delicate linear segments. Most species, including all of those found in our deserts, are completely hairless, but the stems and leaves are often coated with an extremely fine whitish powdery layer. *Eschscholzia* have torpedo-shaped buds protected by two sepals that fall as a unit when the flower opens, exposing the four or six yellow to orange petals. *Eschscholzia* species have 12 to many stamens, and no style. A lobed stigma sits right on top of the tall, thin ovary.

Eschscholzia glyptosperma (desert gold-poppy, Mojave gold-poppy) is easy to spot because all leaves are clustered at the base of the plant with bare flower stems rising above, each supporting a single golden bloom (in contrast to *Eschscholzia californica,* which is distinctly orange). The seeds (*sperma*) are coarsely pitted: "engraved" or "carved" (*glypto-*). Desert gold-poppy is an annual with a preference for sandy washes and open flats up to about 1,500 m. It is somewhat more common in the Mojave Desert than the Sonoran.

Eschscholzia minutiflora (little gold-poppy) is distinctive for being "small" (*minuti-*) "flowered" (*flora*). The yellow petals are generally only about 1 cm long. Leaves are dissected into segments shorter than other desert species, with each segment slightly flared at the tip. Plants have a fairly leafy appearance, although most leaves are near the base. *Eschscholzia minutiflora* can be found in both deserts at altitudes up to about 2,000 m.

Platystemon—This is a single-species genus with a Greek name that means "flat" or "broad" (*platy-*) "stamen" (*stamon*), referring to flat filaments of the stamens.

Platystemon californicus (**cream cups**) is a low-growing (10–30 cm) annual conspicuous for its very long shaggy hairs covering

Eschscholzia minutiflora (little gold-poppy)

leaves, stems, and sepals. It has three sepals and six petals. The petals, 8–16 mm long and almost round, are usually cream-colored, as the common name implies, but they can tend toward yellow. Stamens have flattened filaments and number at least 12, though usually there are more. The unusual ovary has 9–18 stacked segments that split into one-seeded compartments when the fruit matures. Cream cups inhabits the westernmost portions of our deserts where there is grassland and sandy soil, at elevations less than 1,000 m. It is particularly prolific in burned areas.

Examples and Uses

Poppy seeds are well known on bagels and breads and are frequently incorporated into salad dressings, muffins,

Platystemon californicus (cream cups)

and lemon pound cake. Tiny and black, they do not taste like much on their own but somehow manage to enhance the flavor of a wide variety of foods. Poppy seeds and an edible oil pressed from them come from *Papaver somniferum*, the infamous opium poppy.

Opium poppies are the source for illegal drugs such as opium and heroin. The potent painkillers morphine and codeine also are derived from these poppies. These substances come primarily from milky sap that is collected off unripe fruits of the plant. *Papaver somniferum* are among the tallest, brightest, and most spectacular members of the Poppy Family, with flowers reaching 10–12 cm across. In the garden the tall stems (over 1 m) must be staked to keep the heavy flowers from blowing over. Until recently, it was illegal to grow opium poppies, but it is hard to imagine a viable backyard drug operation because the sap is sparse and hard to collect and process. It would be wiser to wait for the seeds and make lemon pound cake.

More common in the garden than the giant opium poppy are Iceland poppies (*Papaver nudicaule*) and several European species of *Papaver*. Prickly poppies and California poppies are important components of low-water landscaping, as is the bush-poppy (*Dendromecon*), also native to southern California but not found in the deserts. The showy matilija poppies (*Romneya*) of our south coast and peninsular ranges do not grow in our deserts. A few poppy genera with bilateral symmetry, such as *Dicentra* and *Corydalis*, are also popular garden plants, requiring a richer, moister soil than desert genera.

Fallugia paradoxa (Apache plume), fruits; Mojave Desert, north side of Clark Mountains

ROSACEAE

ROSE FAMILY

The rose is probably the world's most widely recognized flower and has been incorporated into romantic poetry, literature, and even architecture. The cultivation of roses, pursued for over 3,000 years now, has resulted in the addition of numerous petals (actually enlarged sterile stamens) prized for their abundance, color, curl, size, and iridescence.

Fabulous fruits are another feature of the Rose Family. Apples, in particular, are central to numerous legends, from Eve being tempted in the Garden of Eden to Johnny Appleseed aiding in the dispersal of apples in North America. Other Rose Family fruits include peaches, pears, apricots, cherries, strawberries, blackberries, and raspberries. A close look at their flowers reveals the family resemblance.

The Latin word *rosa,* the root of the family name, means red, pink, or rose and is the name for the genus most typical of the family.

Rosaceae Icon Features

- Radial symmetry with five (or zero) petals and five sepals
- Saucerlike central hypanthium surrounding numerous (sometimes one or a few) separate pistils
- Stamens numerous (10, 15, 20, or more; rarely just five)

Distinctive Features

A saucer-, cup-, or funnel-shaped **hypanthium**—the fused base of sepals, petals, and stamens—is suggestive of the Rose Family, where it is usually a distinctive feature at the center of the flower, with sepals, petals, and stamens all extending separately from it in radial formation. The multiple petals of garden roses is misleading. Our native Rosaceae members have either five separate petals in radial symmetry, or no petals at all. Often, Rosaceae petals are fairly large (1 cm), quite round, and showy. White and pink are the most common colors. Rosaceae flowers also have five sepals (with rare exceptions) and sometimes even five **bractlets**, looking like smaller sepals in between the real ones. Rosaceae species generally have numerous stamens. Sometimes this means only 10, but they can increase in multiples of five until you get tired of counting.

Several genera of Rosaceae have a single pistil

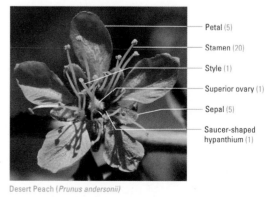

Desert Peach (*Prunus andersonii*)

- Petal (5)
- Stamen (20)
- Style (1)
- Superior ovary (1)
- Sepal (5)
- Saucer-shaped hypanthium (1)

Rose Family (Rosaceae) flowers have most parts in five's, with either a single or multiple pistils.

(stigma, style, and ovary) in each flower and make fruits with a single pit or stone, like plums, peaches, and cherries (*Prunus*). Other genera have four or five or many pistils in a cluster at the center of each flower. The variations on fruits that result from this arrangement include blackberries (*Rubus*), strawberries (*Fragaria*), and rose hips (*Rosa*). Each blackberry and raspberry is made up of numerous tiny single-seeded fruits like a miniprune, arranged over the surface of an elongated receptacle. When you pick raspberries, the receptacle stays on the stem, leaving a cavity in the fruit that children soon notice perfectly fits the tips of their small fingers. With strawberries, it is the receptacle that we eat, along with a multitude of tiny dry fruits that are just dots on the surface of the red receptacle. Rose hips are still another matter, with the hypanthium expanding into a fleshy layer around numerous small pistils. Both the hypanthium and the pistils are attached to the receptacle. When the flower is fresh, the hypanthium surrounds the pistils loosely, like an urn-shaped sleeve. After fertilization, as the fruit matures, the hypanthium expands, forming a round red, pithy casing tightly enclosing the pistils with their ovaries and seeds.

In each of the above examples, the ovary or ovaries are superior. In a few genera of Rosaceae, such as apples (*Malus*) and pears (*Pyrus*), the ovary is inferior. Technically, this results from the hypanthium being completely fused to the ovary so that the two are indistinguishable. Where the hypanthium extends past its fusion with the ovary, it appears to be attached to the top of the ovary, making the ovary inferior to the attachment of other flower parts. Notice that in an apple, the little dried sepals are at the far end of the fruit, away from the stem, whereas strawberries have their sepals at the stem end. Only two desert genera, service-berry (*Amelanchier*) and wild crab apple (*Peraphyllum*), have an inferior ovary like an apple.

Not all Rose Family fruits are good to eat. Several genera produce a small, dry fruit, frequently distinguished by a long plumelike tail—actually the style of the flower, enlarged during development of the fertilized fruit.

Although vegetative features of Rosaceae are extremely variable, leaves almost always have an alternate arrangement on stems and may grow singly or in clusters. The bases of leaves usually have flag- or fringelike stipules. Although compound leaves are common throughout Rosaceae, plenty of species have simple leaves that vary in shape, size, texture, and other features. However, Rosaceae leaves are never fleshy. They may be thickened and hard, but they are not bulbous with watery interiors like many of their close relatives in the Stonecrop Family (Crassulaceae).

Similar Families

Flowers in the Mock Orange Family (Philadelphaceae), a close relative, resemble those of Rosaceae; however, the mock oranges frequently have four rather than five petals. Also, if there are 10 stamens on mock oranges, they alternate in length and are broad and flat near their bases so that the filaments are almost touching where they arise in a neat circle. The easiest indicator of mock oranges is the opposite arrangement of leaves. Even when deciduous leaves are absent, you can tell where leaves were attached by the lumpy little scars left on the stems and by the opposite pattern of much of the branching. One desert Rosaceae species, blackbush (*Coleogyne ramosissima*), has opposite branching and opposite clusters of leaves, but they are small and dark, very different from those in Philadelphaceae.

The Crossosoma Family (Crossosomataceae), also a close relative of roses, is very similar in numerous respects, but the key feature of Rosaceae, the hypanthium, is either missing or very reduced and disklike. Also, Crossosomataceae leaves have either minute or absent stipules, whereas many Rosaceae species have conspicuous stipules.

Although some members of the Buttercup Family (Ranunculaceae) resemble Rosaceae, buttercups are not common in the desert, and the few species found there are not likely candidates for confusion for a variety of reasons. *Delphinium*, the most common of desert buttercups, has bilateral flower symmetry, not found in roses. *Clematis* is a woody vine, a feature not found in desert Rosaceae. The remaining desert buttercup genera all are annuals or perennials, to be compared with only two desert genera of Roasaceae that contain annuals or perennials. These are *Ivesia* and *Potentilla*, both of which have very green sepals and often also bractlets, in contrast to the Ranunculaceae genera, which have petal-like non-green sepals and never any bractlets.

A few other families have both a hypanthium and flower parts—including stamens—in fives. Rosaceae members present in our deserts, except for two species of *Ivesia*, have 10, 15, 20, or more stamens.

Distribution

The 110 genera and 3,000 species of Rosaceae are distributed throughout the world, mainly in temperate climates. The United States and Canada combined have about 62 genera and 870 species. Some of the largest genera are *Potentilla* (cinquefoil), *Prunus* such as cherries and plums, and *Rosa*, which is familiar in its cultivated form all over the Northern Hemisphere.

California has 39 genera of Rosaceae, only three of which are entirely nonnative. Fourteen of California's Rosaceae genera—over one-third of the total—are represented in our deserts, but most of them are small, with only a single species present in the desert, so they are fairly distinctive and easy to learn.

Desert Rosaceae species in California are almost always found at higher elevations, where they get some cool air and a bit of extra water. Species occupying drier sites tend to have adapted with very reduced leaf size, whereas those at better-watered locations may have leaves looking more like those of their domestic relatives.

California Desert Genera and Species

Most desert Rosaceae genera are heavily branched shrubs, but there are herbaceous perennials and an annual as well. Several woody species are armed with prickles, like rose thorns, or with sharp spines that form as the hardened ends of twigs. Although white and pink are typical flower colors for the family, yellow is also well represented in the desert. Flowers vary in size from the large (up to 8 cm across) interior rose (*Rosa woodsii*) to the tiny (2–3 mm) rock spiraea (*Petrophyton caespitosum*). When tiny, Rosaceae flowers are often clustered. Many flowers are an in-between size (1–2 cm across) and strongly resemble familiar apple, cherry, or strawberry flowers.

SIMPLIFIED KEY TO DESERT GENERA OF ROSACEAE

California's desert genera can be easily distinguished from one another on the basis of their growth form, presence or absence of petals, leaf characteristics, and a few features unique to each genus.

1. Annuals and perennials.
   • Growing in rock crevices, often hanging from cliffs (4 spp) .... *Ivesia*
   • Growing at desert mountain lake edges .... *Potentilla biennis*
1' Shrubs and small trees.

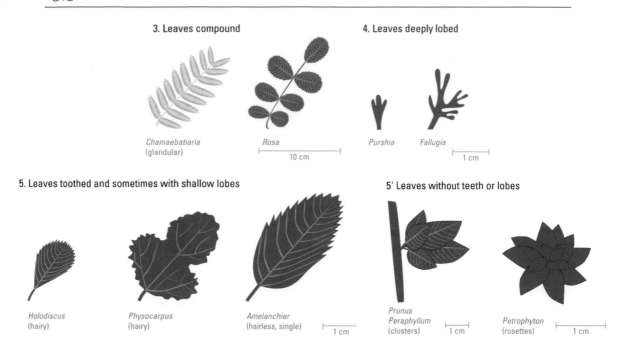

3. Leaves compound

Chamaebatiaria
(glandular)

Rosa

10 cm

4. Leaves deeply lobed

Purshia   *Fallugia*

1 cm

5. Leaves toothed and sometimes with shallow lobes

Holodiscus
(hairy)

Physocarpus
(hairy)

Amelanchier
(hairless, single)

1 cm

5' Leaves without teeth or lobes

Prunus
Peraphyllum
(clusters)

1 cm

Petrophyton
(rosettes)

1 cm

Rose Family (Rosaceae) leaf shapes are used in the key to help determine genus.

2. Flowers without petals, but sepals may be yellow.
  • Flowers with 4 yellow sepals looking like petals. Opposite branching .. *Coleogyne ramosissima*
  • Flowers with 5 tiny sepals. Alternate branching (2 spp) .... *Cercocarpus*
2' Flowers with 5 petals and 5 sepals.
  3. Leaves compound.
    • Leaflets larger than 1 cm, flowers with large pink petals .... *Rosa woodsii*
    • Leaflets numerous and tiny, fernlike, with strong spicy odor ... *Chamaebatiaria millefolium*
  3' Leaves simple.
    4. Leaves deeply lobed like fingers to the palm of the hand, and small (about 1 cm).
      • Petals white, more than 20 pistils .... *Fallugia paradoxa*
      • Petals creamy white to yellow, 1–5 pistils (2 spp) .... *Purshia*
    4' Leaves without deep lobes but with or without teeth.
      5. Leaves toothed and sometimes with shallow lobes (3–5 broad lobes also toothed).
        • Flowers large (petals 1 cm). Ovary inferior. Fruit like a tiny apple *Amelanchier utahensis*
        • Flowers small, clustered along tips of branches. Ovaries 5, superior ....
        .... *Holodiscus microphyllus*
        • Flowers small, bundeled at leaf axils. Ovary superior .... *Physocarpus alternans*
      5' Leaves without teeth or lobes.
        • Plant only a few cm tall, spreading over the surface of rocks ..... *Petrophyton caespitosum*
        • Plant knee-high to tall. Ovary superior. Fruit like a mini-peach (3 spp) .... *Prunus*
        • Plant 1–3 m. Ovary inferior. Fruit like a mini-apple .... *Peraphyllum ramosissimum*

Amelanchier utahensis (Utah service berry)

Cercocarpus ledifolius (curl-leaf mountain-mahogany), fruits

Amelanchier (**service-berry**)—The Latin name for this genus is derived from an old French word for a similar plant in the same family. Only about 10 species are found in various parts of the world, with two native in California. The genus is distinctive for having blue-black berry-sized fruits that arc constructed similar to an apple. The common name may refer to its usefulness to Native Americans and early settlers as a food long before serious apples were brought to the West.

Amelanchier utahensis (**Utah service-berry**), named for another state, is the only species growing in the California deserts, where it is confined to the somewhat cooler conditions of mountainous habitats. It is a shrub, usually about as tall as a person, with fairly large (up to 4 cm) oval, deciduous leaves that generally have small teeth at the tip end. White flower petals are about 1 cm long and less than half as wide. They generally look loosely wavy rather than regimental. In addition to the inferior ovary, look for 10–20 stamens.

Cercocarpus (**mountain-mahogany**)—This is a large shrub or small tree with small (usually 1–2 cm) leathery, evergreen leaves with rolled-under edges. Like service-berry, it grows at higher locations in California's desert mountains. The common name refers to the quality and location of the wood, which can grow to a fairly sub-stantial trunk. The Latin name means there is a "tail" (*cerco-*) on the "fruit" (*carpus*), and it is mostly the tail that you will see. It is actually the style of the flower that elongates as the fruit develops, becoming a silvery-white plume. Dozens of them spread from the small branches where flowers have been abundant. Flowers, having no petals, are much less noticeable than mature fruits. Each tiny flower has 10–15 or more stamens but just one pistil with a superior ovary, a single ovule, and a single style that outsizes everything else as the fruit develops.

Cercocarpus ledifolius (**curl-leaf mountain-mahogany**) can reach 10 m in height. The leathery leaves may be up to 4 cm long but usually are half that size with curled-under edges. The specific epithet describes the "leaf form" (*folius*) as resembling the leaves of *Ledus glandulosa*, Labrador tea, in the Heath Family (Ericaceae). The variety found in the desert mountains is known as *intermontanus*, meaning "between the mountains": that is, between the Rockies and the Sierra Nevada. This variety is notable for the length (up to 7 cm) of mature styles.

Chamaebatiaria (**fern bush, desert sweet**)—This is a single-species genus similar to *Chamaebatia*, a nondesert genus whose Latin name means "low" (*chamae-*)

Chamaebatiaria millefolium (desert sweet)

Colyogyne ramosissima (blackbush)

"bramble" (*batia*), neither of which describes our unarmed shrub that stands 1–2 m tall. Both common names are better descriptors: the finely divided leaves lead many people to think they are looking at a fern, and the name "desert sweet" refers to the strong odor of the glandular foliage. In our desert Rosaceae, only the *Chamaebatiaria* has strongly aromatic fernlike leaves.

Chamaebatiaria millefolium (**desert sweet**) has a Latin specific epithet that refers to its "thousands" (*mille-*) of tiny "leaves" (*folium*)—actually the secondary leaflets of the twice-compound leaves. Each leaf is 2–8 cm long with 31–43 primary leaflets, each of which is again divided into 11–35 secondary leaflets, which could indeed account for as many as 1,505 leaflets (43 × 35) on a single leaf. There are also numerous small white flowers (with petals 5 mm long) clustered along the ends of branches. Each flower has numerous stamens and four to five partially fused superior ovaries. Once again, this is a shrub of the mountainous portions of the deserts.

Coleogyne (**blackbush**)—The common name of this single-species genus refers to the dark coloring of the numerous branches and small clustered leaves. The Latin name says the "female parts" (*-gyne*) are in a "sheath" (*cole-*), referring to the superior ovary being surrounded by a part of the hypanthium.

Coleogyne ramosissima (**blackbush**) is a stiff and spiny, intensely branched (*ramos-*, or "branched," plus *-issima*, "the most") dark shrub, usually under a meter in height. A good coating of yellow flowers (the sepals) lightens up this member of many dry slope communities in spring. There are 30–40 stamens and only one pistil with a very hairy style. The most distinctive feature of the petal-less flower is the four sepals (instead of five) that look like yellow petals. Another inconsistency with the rest of the family is that leaf bundles and branches are arranged opposite one another rather than alternately. Blackbush grows between 600 and 1,600 m in the desert mountains.

Fallugia (**Apache plume**)—This genus is yet another single-species genus in the Rose Family, growing as a shrub in our desert mountains. It is named after the seventeenth-century Italian abbot and botanist Virigilio Fallugi.

Fallugia paradoxa (**Apache plume**) may seem paradoxical (*paradoxa*) because of the many large white blooms on a dry desert shrub, but *Fallugia paradoxa*, with its many pistils and many stamens, is a typical Rose Family member. Each of the pistils develops an elongated (3–5 cm) plumelike style

Fallugia paradoxa (Apache plume), flower

Fallugia paradoxa (Apache plume), fruits from a single flower

as the flower fades and the fruits mature. These styles, usually a couple of dozen per flower, linger for a long time, easily revealing the identity of this shrub when the flowers are gone. Possibly the plume resembled a feature of an Apache headdress, but when low sun catches a bush covered with plumes, it lights up like a brush fire. Leaves of *Fallugia* are similar to those of *Purshia*, but the most notable distinction between the two genera is the abundance of pistils in flowers of *Fallugia*, which can have dozens, in contrast to the one to several pistils of *Purshia* species.

Holodiscus (**cream bush**)—This genus has five species, all in the Western Hemisphere. Two are in California, and one variety of one species is in the desert mountains. The Greek genus name, meaning "whole" (*holo-*) "disk" (*discus*), refers to the hypanthium, which is not lobed. There is also a nectary disk (appearing as a swollen band near the rim of the saucer-shaped hypanthium) that produces the attractant for pollinators. Each flower has five petals, five sepals, five pistils, and 15–20 stamens. More notably, the small (less than 1 cm) creamy white flowers are clustered at the ends of branches, very much like *Chamaebatiaria millefolium*.

Holodiscus microphyllus var. *microphyllus* (**rock spiraea**) is Greek for "small" (*micro-*) "leaved" (*phyllus*). Actually, leaves on this species are of different types depending on whether they are crowded

Holodiscus microphyllus var. *microphyllus* (rock spiraea)

along short peglike shoots or spaced along a regular branch. In either case, leaves are less than 3 cm long, and they may be variously toothed and hairy on one or both surfaces. *Holodiscus microphyllus* leaves in no way resemble the fernlike leaves of *Chamaebatiaria millefolium*. Just remember "small leaved" versus "thousands of leaves (leaflets)" to avoid confusion because of the similarity of flowers in these two species. Rock spiraea grows in rocky crevices up to 4,000 m.

Ivesia—Lieutenant Eli Ives (1779–1861) led one of the Pacific Railroad Surveys to the American West and was later a pharmacologist at Yale University. There are 30 species of *Ivesia* in western North America, 22 of which occur in California. In general, *Ivesia* are glandular perennial plants with finely divided leaves growing mostly at the base and greatly reduced in size and number along the rising flower stalk. Usually, leaflets are very small and grow around a central axis, so the entire leaf is long and cylindrical like a furry tail. Some species, however, including all those in our deserts, have flat leaves with only a few large leaflets. Flowers usually have five small bractlets below five sepals, five yellow petals, and 10 or 20 stamens. There may be one to numerous pistils, which have an odd feature: the style is joined to the ovary a bit below its tip, as if the style slid to one side rather than growing as an upright extension from the ovary. Although many *Ivesia* grow erect in meadowlike habitats, all four California desert species grow as hanging plants in rock crevices.

Some exceptions to *Ivesia* flower features occur in our desert species. *Ivesia arizonica* var. *arizonica* generally has no bractlets and only five stamens. It is uncommon, inhabiting limestone crevices of northern desert mountains. *Ivesia patellifera*, which is rare and found in granite crevices of eastern desert mountains, also is usually without bractlets and has either five or 10 stamens.

Ivesia saxosa (rock five-finger) is the most common of our desert *Ivesia* species. It is named "of the rocks" (*saxosa*) for its home in granitic and volcanic crevices at elevations from 900 to 3,000 m. Leaves are flat with up to nine somewhat round leaflets, spaced well apart and variously toothed or slightly lobed. The 2–4 mm yellow petals are a bit smaller than the green

Ivesia saxosa (rock five-finger)

sepals, and the bractlets are smaller still. There are 15–40 stamens and 3–20 pistils.

Peraphyllum (wild crab apple)—Again we have a single-species genus of shrubs, this one named "completely" or "thoroughly" (*per-*) "leafy" (*phyllum*).

Peraphyllum ramosissimum (wild crab apple) is "the most"

Peraphyllum ramosissimum (wild crab apple)

(*-issimum*) "branched" (*ramos-*). In addition to lots of leaves and branches, this shrub has numerous white to pink (or rose) flowers and fruits, very similar to apples but much smaller (1 cm) than normal eating apples. Leaves come in small clusters and are deciduous. Wild crab apple occurs in our northern desert mountains and points north.

Petrophyton (**rock spiraea**)—This genus, named "rock" (*petro-*) "plant" (*phyton*) in Greek, has four species in western North America, only one of which grows in California, where it occurs on limestone of the desert mountains. Note that the common name for this genus is the same as that applied to the totally different looking *Holodiscus microphyllus* var. *microphyllus,* reminding us that there is a sensible reason for scientific naming. The Greek "rock plant" best describes this genus, which is matted right over the surface of rocks, almost looking like a part of the rock but with a different texture. Members of the genus are actually evergreen shrubs, with tiny leaves so tightly packed as to form a stiff surface.

Petrophyton caespitosum **ssp.** *caespitosum* (**turf spiraea**) grows in "turflike patches" or "tufts" (*caespitosum*), referring to the dense clusters of leaves packed across the face of the rocky substrate. Even with leaves that are only a few millimeters long, a single plant can cover almost a meter of rock and have dozens of flower spikes. Each tiny flower has five white to cream petals, five sepals, five pistils, and many stamens, but it is the spike of numerous flowers extending about 10 cm above the mat of leaves that is the main feature.

Prunus—This genus, named after the plum (*prunus* in Latin), includes most of our favorite Rose Family fruits: cherries, apricots, peaches, and others with a single pit or stone covered with soft flesh

Petrophyton caespitosum **spp.** *caespitosum* (**turf spiraea**)

(a fruit type known to botanists as a drupe). There are about 400 species and probably hundreds of cultivars, developed for varying tastes and horticultural conditions. Only seven species are native to California, and three of these occur in our deserts.

Prunus andersonii (**desert peach**) fruits look like dollhouse-sized peaches, but it is the flowers that are special, bright pinkish-red and tightly packed onto the well-armed shrub. Flowering shrubs stand out at a distance on rocky slopes and flats at 900–

Prunus andersonii (desert peach)

Prunus fasciculata var. *fasciculata* (desert almond), functionally male flowers with plenty of pollen

2,600 m. In our deserts they grow only in the most northerly mountains, but they are common throughout the eastern Sierra Nevada and north into the Great Basin. The numerous small twigs harden and become long spines. Leaves are 3 cm or less long, oval, and clustered. The namesake for the species is Charles Lewis Anderson (1827–1910), a physician and a naturalist residing in California and Nevada much of his life.

Prunus fasciculata var. *fasciculata* (**desert almond**) is another spine-covered deciduous shrub. *Fasciculata*, or "fascicled," refers to the clustered arrangement of the small (7–15 mm) leaves, a growth pattern similar to that of *Prunus andersonii*. Flowers of *Prunus fasciculata* are small, with 2–4 mm petals that are white or creamy yellow. Flowers are often exceedingly numerous and exude a musty fragrance that attracts honeybees

Prunus fasciculata var. *fasciculata* (desert almond), functionally female flowers, forming fruits

from miles around. Al-though the flowers look similar, many populations have the sexes on different plants. Flowers on male plants have highly developed stamens and a tiny ovary, while flowers on female plants do the reverse. Tiny fruits are somewhat almond shaped and fuzzy. Desert almond is widespread throughout California's deserts at elevations of 700–2,200 m.

Purshia (**antelope bush**)—Frederick T. Pursh (1774-1820) was a botanical explorer who published a flora of North America in 1817. This genus comprises five species, all located in western North America; two of them occur in Cali-fornia and are present in the deserts. *Purshia* species are shrubs or small trees with small (about 1 cm) deeply lobed leaves that have stiff rolled-under edges and may have embedded glands that look like tiny pinheads sunk-en into the upper surface of the leaf. Flowers have about 25 stamens and one to several pistils that mature to fruits with long plumose styles. Although these fruits are similar to those of *Fallugia paradoxa*, on *Purshia* each flower produces only one to a few, whereas *Fallugia* flow-ers each generate dozens.

Purshia mexicana var. *stans-buryana* (**cliff rose**) was named *Cowania stansbur-iana* until recently, and retains the varietal name honoring Captain Howard Stansbury (1806–1863), leader of a government ex-pedition to explore the Great Salt Lake Basin in 1850. The specific epithet, *mexicana*, refers to the occurrence of

Purshia mexicana var. *stansburyana* (cliff rose)

the species in Mexico as well as the southwestern United States. The leaves of this plant generally have five to nine lobes, looking a bit like the outstretched fingers of a tiny hand. White or cream petals are up to 8 mm long, making showy flowers that may totally cover the bush. There are usually four to five pistils, with styles that become 2–6 cm long and plumose as fruits mature. *Purshia mexicana* var. *stansburyana* is found in desert mountains between 1,100 and 2,500 m.

Rosa (**rose**)—Finally, the genus that supplies the family name is the last (alphabetically) to provide a California desert species. Although California has nine native species of *Rosa*, only one grows in the deserts, where it is re-stricted to moist locations such as stream banks or seeps. *Rosa* species may be shrubs or vines with prickles, which we recognize as thorns, and pinnately compound leaves with conspicuous stipules, like flags. Native members of the genus usually have five pink petals, five long pointed or toothed sepals, many stamens, and many pistils.

Rosa woodsii var. *ultramontana* (**interior rose**) is the lovely pink-flowered rose that we find growing in dense thickets in moist desert mountain habitats. *Ultramontana*, meaning "across the mountains," refers to its "interior" location east of the Sierra Nevada. Its petals, at 15–20 mm, are larger than those of any other desert species of the family. By fall, mature fruits are bright orange-red, round, and about 1 cm across. In addition to vitamin C, they contain pectin and can be used to make jelly or to help jell the jellies of other fruits.

Rosa woodsii var. *ultramontana* (interior rose)

Examples and Uses

Examples of Rosaceae abound in our neighborhoods and grocery stores as well as in the deserts. Roses are certainly the most familiar example of the Rose Family, but several other genera are well known as landscape plants. These include *Cotoneaster*, which is usually evergreen and may be used as groundcover or shrubbery; *Pyracantha*, which has abundant orange-red berries in the fall; and *Crataegus*, the hawthorns, which are small flowering trees.

Many familiar fruits are from the Rose Family; however, only strawberries and blackberries have native American ancestors. Plums are from the Caucasus. Apricots and peaches originally came from China, though California now is the major supplier worldwide. Apples, cherries, and pears originated in Europe and western Asia.

In addition to the fleshy fruits of Rosaceae, we have almonds, which are the seeds of fruits from a tree similar to a peach. In general, Rosaceae seeds are poisonous, due to their release of cyanide when digested. Commercial almonds have been bred to have low cyanide levels, but still, large quantitites should not be eaten green. Except for the berries with tiny seeds, all Rosaceae seeds should be avoided. The occasional pear or apple seed is generally harmless, but a handful could be fatal.

A few medicinal uses of Rosaceae have benefited humans, including the use of rose hips as a source of vitamin C, which goes back to ancient Native American practices. More recently, an ingredient of apricot seeds known as laetrile has been found to be useful in the treatment of some cancers.

The wood of certain Rosaceae species is used for fine cabinetry and furniture. Cherry wood, for example, prized for its warm reddish tone, comes from a forest species of eastern North America. Other fruit woods, often from the cultivated species, also are distinctive and desirable. Rosewood, familiar to us as a dark, rosy, and lustrous wood often used in Scandinavian-style furniture, is not from the Rose Family; it is a Pea Family (Fabaceae) member of tropical origin.

Finally, the excellent aroma of some roses has led to their use in perfumes and cuisine.

Castilleja angustifolia (desert paintbrush), northwest of Rainbow Basin Natural Area

SCROPHULARIACEAE

FIGWORT FAMILY

The Figwort Family is large and diverse, even including some parasites sucking nutrients from the roots of other plants. You will not suspect this subsurface subterfuge, however, because the parasitic genera also have green photosynthetic tissue and look respectable, like other currently accepted members of the family (Jepson manuals, 1996 and 2002). However, big changes are in store for Scrophulariaceae, which is now believed to include many species with closer ties to other families. Species with parasitic capabilities are likely to be moved to Orobanchaceae (Broom-Rape Family), while many others, including your garden favorites, will likely join Plantaginaceae (Plantain Family).

The root of the family name, the Latin word *scrofula*, describes a tumor or swelling condition that was supposed to be cured by some herbs in the genus *Scrophularia*. "Wort" is an archaic name for a plant with medicinal qualities, while "fig" refers to a disease, *ficus*—better known as hemorrhoids—which figwort was said to cure. In addition to figworts, the family contains louseworts (*Pedicularis*) and mudworts (*Limosella*). It might be more elevating to think of the family as foxgloves (*Digitalis*) or snapdragons (*Antirrhinum*), and make up your own reasons for the common names. Our preference is to call them all "scrophs."

Scrophulariaceae Icon Features

- Tubular corolla with bilateral symmetry
- Staminode common, usually like a hairy tongue lying along the floor of the corolla
- Four stamens (sometimes two), usually curved up to position anthers above visiting insects

Distinctive Features of Scrophulariaceae

Look for a two-lipped tubular corolla and a superior ovary, ovoid in shape and containing many (or rarely only six) ovules. There is one style, but the stigma is generally two-lobed, reflecting two chambers in the ovary. There are sometimes two, but usually four, stamens, and sometimes also one (rarely two) **staminodes**, or infertile stamens that look very different from the fertile ones and lack an anther. Flower colors are highly variable but often brilliant. Flowers may grow in elongated clusters at the tips of stems or singly or paired in leaf axils.

Plants may be annuals, perennials, subshrubs, or shrubs; this family contains no trees or plants with extensive amounts of woody tissue. Many of our desert genera have an opposite leaf arrangement, with each pair of leaves turned perpendicular to the pair directly below. With only one exception, however, this does not result in a square stem, as is predominant in the Sage Family (Lamiaceae); rather, Scrophulariaceae stems are almost always round in cross section. For the family as a whole, alternately arranged leaves are more common than opposite, and several genera in our deserts fit this pattern. Leaves are generally some sort of oval shape, with or without teeth, but usually not deeply lobed.

Similar Families

If it sounds hard to pin down Scrophulariaceae, it is. The family is highly variable, and probably will be split into several different families before long. Furthermore, the desert is full of close relatives. The largest desert family with similar flowers is the Sage Family (Lamiaceae), but several fairly reliable simple differences can be pointed to. Sages usually have stems that are square in cross section, but there is only one desert scroph with a square stem (*Scrophularia*). Sages are usually strongly aromatic, with familiar odors like mint or lavender. Scrophs do not share this feature, and you will not find them in your spice rack. Sages often have dense spherical clusters of flowers at each of several upper bracts or leaf pairs, while Scrophulariaceae flower clusters tend to be sparse or evenly distributed along the tips of stems. For the definitive analysis, look at the fruit: sages have a deeply four-lobed fruit, each lobe with a single seed, while scroph fruits are rarely lobed, and certainly not deeply four-lobed, and they usually have more seeds than you'd care to count.

The Acanthus Family (Acanthaceae) has many features in common with scrophs. Chuparosa (*Justicia californica*), which is the only member of the family common in California, has bright red two-lipped, tubular flowers and opposite branching and leaves, but again, the fruits distinguish it clearly from scrophs. Acanthus Family fruits are specialized for flinging seeds away from the parent plant. Look for dried fruits, opened in a V-shape, with little launching platforms for each of two seeds (now long gone) on each side of the catapult.

The treelike form and extensive woody branching of desert willow (*Chilopsis linearis*), the only desert member of the Bignonia Family (Bignoniaceae) native to California, rule it out as a Scrophulariaceae representative, even though the flower structures of the two families are very similar. The same is true for several Olive Family (Oleaceae) members, while other Oleaceae and Buddlejaceae (Butterfly Bush Family) species have unisexual flowers that will not be found among the scrophs (at least until a family reorgnization is accepted).

The closely related Broom-Rape Family (Orobanchaceae) is a root parasite that somewhat resembles paintbrush (*Castilleja*), in Scrophulariaceae. However, *Castilleja* has plenty of green tissue, looking "normal" in comparison to the nongreen pigmentation of *Orobanche*. However, *Castilleja* and other semiparasitic genera now in Scrophulariaceae may soon be considered part of Orobanchaceae.

One more small family, Unicorn-Plant Family (Martyniaceae), has corollas meeting the general description of scrophs, but its huge (5–10 cm), woody fruits readily distinguished it.

Except for the Sage Family (Lamiaceae), families resembling Scrophulariaceae have very few desert representatives in California and are easily recognized. Learn to eliminate sages quickly, memorize chuparosa, excommunicate trees and nongreen plants, avoid huge woody fruits, and then feel fairly confident that scrophlike flowers are, in fact, members of Scrophulariaceae—at least for now.

Family Size and Distribution

Scrophulariaceae has around 200 genera and 3,000 species distributed throughout the world, with a preference for north temperate areas. The largest genera, each with over 300 species, are somewhat poorly represented in North America; they include *Pedicularis* (louseworts), *Calceolaria* (slipper-flower), and *Verbascum*. The United States and Canada combined have about 70 genera and over 850 species, with the greatest representation in *Penstemon* (beardtongue), *Castilleja* (paintbrush), and *Mimulus* (monkey flower).

California has 32 genera of Scrophulariaceae, 13 of which have desert representatives—the largest genera in both cases likewise being *Penstamon, Mimulus,* and *Castilleja*. One-quarter of the state's scrophs are exclusively nonnative, but none of these live in California's deserts.

Scrophulariaceae species fill virtually every habitat type of our deserts from dry sandy washes to rocky crevices. There are species for running water and for gravel slopes at every elevation. Species diversity is greatest toward the north and toward higher elevations, as might be expected from the north temperate preference of the family as a whole.

California Desert Genera and Species

Brightly colored tubular, two-lipped flowers are likely to be Scrophulariaceae. Common colors are yellow, red, magenta, and blue to purple, and flowers generally are 1–4 cm in length. Most plants are big enough to spot easily when in flower. Look for leaves that are relatively simple in outline, often with teeth. In the deserts only *Castilleja* (paintbrush) and *Cordylanthus* (bird's beak) have leaves that may be deeply three- to five-lobed, and these tend to grade into bracts that are similarly lobed and usually pigmented as if they were corollas.

SIMPLIFIED KEY TO DESERT GENERA OF SCROPHULARIACEAE

Leaf arrangement, either predominantly alternate or opposite, makes a good starting point for distinguishing desert genera of Scrophulariaceae. Leaves within the family have differing arrangements of their major veins, either palmate or pinnate. Details of corolla shape are important, especially the tubular part, which may have a pocket or spur on one side. Numbers of stamens, either two or four, and features of the single staminode, if present, are especially useful for determination of species as well as genus.

1. Leaves alternate, at least the upper leaves. Lower leaves sometimes opposite or rarely all basal.
  2. Corolla base with a pocket- or cuplike bulge on one side. Corolla lips generally expanded and flared outward.
    3. Leaves palmately veined, either triangular with no teeth or circular with sharp teeth (2 spp) .. .. *Maurandya*
    3' Leaves pinnately veined and entire (no teeth or lobes).
      • Corolla more than 15 mm long, cream or yellow, spotted. Plants compact, leafy, hairy annuals (2 spp) .... **Mohavea**
      • Corolla less than 14 mm long, white, cream, or yellow, often with lavender veins. Plants often twining, typically without hairs (4 spp) .... **Antirrhinum**
  2' Corolla base not with a pocket- or cuplike bulge, just a generally symmetrical tube. Corolla lips narrow, straight, and pointed like a robin's beak, or else tiny.
    • Corolla upper lip much longer than lower. Corolla, calyx, and bracts often colorful, looking like a bright red to magenta paintbrush. Plants mostly perennial, with a few upright branches from near the base (6 spp) .... **Castilleja**
    • Corolla lips of about equal length. Corolla, calyx, and bracts not brightly colored, usually marked with maroon or dark red. Annuals, branched throughout (3 spp) .... *Cordylanthus*
1' Leaves opposite or sometimes whorled.
  4. Stamens 2. Corolla appearing 4-lobed (upper 2 lobes fused), blue or blue-violet (2 spp) .... *Veronica*

4' Stamens 4, with or without 1 staminode. Corolla generally 2-lipped, variously colored.

  5. Calyx with distinct tubular base.

    • Corollas yellow, orange, red, magenta, or pinkish white. Calyx prominently ribbed or pleated, sometimes upper lobes larger than lower (15 spp) .... *Mimulus*

    • Corollas lavender to purplish or lavender-blue. Calyx without distinctive ribbing or pleating, 5 almost equal lobes (2 spp) .... *Collinsia*

  5' Calyx with little or no tubular base.

    6. Staminode absent.

      • Corolla white to pink, 5 nearly equal spreading lobes .... *Bacopa monnieri*

      • Corolla violet to purple, upper lip 2-lobed and arched over the mouth of the tube, tube almost square in cross section .... *Stemodia durantifolia*

    6' Staminode present, sometimes glandlike or hairy and very different from stamen filaments.

      7. Stem square in cross section .... *Scrophularia desertorum*

      7' Stem round in cross section.

        • Fertile stamens with hairy filament bases. Corolla yellow or brownish yellow to cream with darker lines. Upper corolla lip cupped like a hood arching forward (2 spp) ....
        .... *Keckiella*

        • Fertile stamens with hair-free filament bases. Corolla pink to bluish, purple. Upper corolla lip flared outward, not cupped or hooded (19 spp) .... *Penstemon*

Antirrhinum (**snapdragon**)—Grandmothers know that if you squeeze on the sides of the corolla tube of a snapdragon, the mouth of this dragon will snap open and closed. The Greek name, *anti-* (against or opposite) *-rhinum* (a nose or snout), is usually interpreted as meaning "noselike," possibly for the bulge at the base of the corolla tube. There are 36 species of snapdragons, mostly in western North America and the Mediterranean. California has 15 species, a few of which are nonnatives. Four species inhabit our deserts, all of which are native. *Antirrhinum* are annuals, often climbing like vines, and have leaves that may be opposite toward the base of the plant but alternate above.

Antirrhinum filipes (**twining snapdragon**) is called "threadlike" (*filipes*) for the extremely long (up to 10 cm), slender, and twining pedicels that support each flower. Corollas are about 1 cm long and yellow, sometimes with maroon spots on the lower lip. The twining stems climb high into shrubs, but the plants are so thin that they may be hard to spot. *Antirrhinum filipes* is the most common of our desert species and can be found bordering washes under 1,400 m, where it climbs a convenient shrub or covers piled debris.

Antirrhinum filipes (twining snapdragon)

Castilleja (paintbrush, owl's clover)—This genus of around 200 species is named after the Spanish botanist Domingo Castillejo (1744–1793). One of its common names reflects the definite resemblance to an artist's paintbrush held brush up, dripping with bright red or magenta paint (although not all species have these brilliant hues). Colorful parts of the plant include mainly the calyx and bracts more than the corolla, which is hardly visible. The calyx usually has four lobes, and the depth of separation between lobes is important in using most species keys. Every flower has its own colorful bract, and below the flowering portion of the stem, deeply lobed leaves often mimic the bracts. California has 37 species of *Castilleja*, with six occurring in our deserts.

Castilleja angustifolia (desert paintbrush) will not go unnoticed, appearing as a brilliant splash of red visible in the desert from a considerable distance. Closer investigation reveals a gray-green perennial, each erect branch of which terminates in a cluster of colorful

Castilleja angustifolia (desert paintbrush)

flowers and bracts. The specific epithet says it is a "narrow" (*angusti-*) "leaved" (*folia*) species, and each leaf may have up to five lobes. *Castilleja angustifolia* grows mainly in the Mojave Desert and points north, preferring elevations of 1,000–3,000 m.

Castilleja exserta (purple owl's clover) is common in the western Mojave Desert under 1,600 m. Its brilliant magenta coloring will attract your attention, and on close inspection you will see tiny white or yellow markings on the corolla that look like an owl's face. *Exserta*, meaning "protruding," refers to the tiny but colorful corolla protruding above a similarly colored bract. Purple owl's clover is an annual, usually only 20–30 cm tall, but it can be very abundant and dramatic in wet years.

Castilleja exserta (purple owl's clover)

Castilleja foliolosa (woolly paintbrush)

Castilleja foliolosa (**woolly paintbrush**) is a heavily branched subshrub or shrub, densely covered with branched hairs that make it appear "woolly." *Foliolosa* means "having leaflets," referring to leaves with almost separate linear lobes. Calyces and bracts are brilliant red, making this shrub obvious when it flowers. It is found on the western edges of our deserts.

Keckiella (**bush beardtongue**)— The California botanist David D. Keck (1903–1995) is honored in the genus name for this group of seven species of shrubs and sub-shrubs, two of which occur in our deserts. Stems tend to be long and leafy, with pairs or triplets of leaves and one to few flowers growing from leaf axils near the tips of branches. Leaves are usually only about 1 cm long and fall off during extended dry periods. Corollas tend to have an upper lip that is cupped like a hood. The four stamens have filaments with dense hairs at the base. The staminode may or may not be hairy at the tip or at the base. The common name, beardtongue, applies also to the genus *Penstemon*, in which the staminode, or "tongue," is almost always hairy or "bearded." Referring to *Keckiella* as bush beardtongue helps to distinguish these plants from *Penstemon,* which are usually perennials, although the distinction made in the common name is not always adequate.

Keckiella antirrhinoides (**yellow bush beardtongue**) has yellow flowers reminiscent of *Antirrhinum filipies*—as the specific epithet implies. *Keckiella antirrhinoides* corollas are distinctly hooded and hairy, and the small opposite or whorled leaves on these shrubs make identification easy. It grows in both deserts at elevations from 100 to 1,600 m but is seldom common.

Mimulus (**monkeyflower**)—This genus of at least 100 species is widely distributed in western portions of the Americas as well as several other parts of the world. Its Latin name means "little mime," referring to the comic facelike markings of some of the corolla lobes. Plant form is highly variable, including annuals (mostly) to shrubs, erect (mostly) to spreading, hairy or not, glandular or not, and flowers of almost any color. Leaves are reliably opposite, and they are usually some sort of an oval shape, frequently with teeth. Corollas have a tube with two lengthwise folds along part of the lower side, and sometimes the base of the lower lip is raised in a mound that obscures the opening of the tube. There are four stamens but no staminode. The calyx usually is distinctive for being mostly tube with five small lobes formed by the extended points of pleats in the tube.

Keckiella antirrhinoides (yellow bush beardtongue)

Sometimes upper lobes are larger than lower ones. Features of the calyx and corolla, corolla color, and length of the flower pedicel are important in identifying to species using keys, which you will definitely need for at least some species. California has 63 species of *Mimulus*, 15 of which occur in our deserts. Several of our desert species are very distinctive and easy to recognize, though they are not always easy to find.

Mimulus aurantiacus (bush monkeyflower)

Mimulus aurantiacus (bush monkeyflower) is the only shrub or subshrub of the genus that occurs in California. It is common in the state, but in our deserts it is restricted to the edges, usually on rocky slopes. Corollas are usually orange (*aurantiacus*) but may range from white to red.

Mimulus bigelovii (Bigelow monkeyflower) is a common annual with a brilliant magenta corolla spotted with yellow near the tube. Sometimes the magenta pigmentation is so exuberant that the stems, leaves, and calyces are all slightly tinted. It is extremely hairy and usually only grows about 10–20 cm tall, but if it is flowering you will see it. *Mimulus bigelovii* was named for Dr. Jacob M. Bigelow (1787–1879), a professor at Harvard Medical School and author of *American Medical Botany*. The *Mimulus* bearing his name likes sandy washes, gravel slopes, and little-used roadbeds up to 2,300 m.

Mimulus cardinalis (cardinal monkeyflower) is named for its cardinal-red coloring, unusual among *Mimulus*. Usually a tall hairy perennial that spreads readily under the right conditions, it can grow to over 1 m in height or it can flower when no more than knee high. It likes running water, or at least

Mimulus bigelovii (Bigelow monkeyflower)

Mimulus cardinalis (cardinal monkeyflower)

Mimulus guttatus (yellow monkeyflower)

Mimulus rupicola (Death Valley monkeyflower)

spots where that event is fairly reliable, up to about 2,400 m. Hummingbirds are very territorial over these nectar-bearing red flowers, often patrolling their stream segments aggressively.

Mimulus guttatus (**yellow monkeyflower**) is a common water-loving annual with lots of yellow flowers. It can live longer than one year and become a perennial if conditions are favorable, but the seasonal dryness of most desert habitat means that it must reproduce by seed. It has hollow stems and leaf blades that are attached directly to the stem without the benefit of a petiole. *Guttatus* refers to the "speckled" or "spotted" condition of corollas.

Mimulus rupicola (**Death Valley monkeyflower, rock midget**) is uncommon and occurs only in limestone crevices of a few mountain ranges bordering Death Valley. *Rupi-* ("rock") *cola* ("dweller") refers to this habitat requirement. Plants are small (1–17 cm) and perennial, with thin leaves. Corollas are pale pink with a magenta to purple spot at the base of each lobe and some yellow in the tube.

Mohavea—This genus—named for the Mojave River, where it was first collected—has only two species, both of which occur in California's deserts and nearby. They are hairy annuals with a lot of leaves and with a single flower in each upper leaf axil. There are just two fertile stamens, plus two staminodes. Since most Scrophulariaceae genera have either no staminodes or only one, this is a good distinguishing feature.

Mohavea confertiflora (**ghost flower**) bears this common name because the flower is so translucent and pale yellowish that it almost is not there. At first glance, flowers might resemble those of *Mentzelia involucrata* (Loasaceae), but the similarity is due to flower color only. *Mentzelia* flowers have radial symmetry and separate petals with no tubular portion at all, whereas *Mohavea* has the usual scroph bilateral symmetry and a corolla with a tubular base. *Confertiflora* means that the "flowers" (*flora*) are "dense" or "crowded together" (*conferti-*). Ghost flower occurs up to 1,100 m in both deserts south of Inyo County. Lesser Mohavea (*Mohavea brevifolia*), which grows in many parts of Inyo County, including Death Valley, has yellow flowers and is a little smaller than ghost flower, but they are easily recognizable as first cousins.

Penstemon (**beardtongue**)—This is the largest genus of Scrophulariaceae in North America, and all 250 species are native only to this continent. California has 53 species of *Penstemon*, and the deserts host 19 of these, almost always in the mountains. The genus name refers to the staminode as "almost" (*pen-*) a "stamen" (*stemon*, also meaning "thread"). The name beardtongue, too, refers to the staminode, which is positioned like a tongue and usually is hairy like a beard. *Penstemon* species are perennials to shrubs with opposite leaves that are often toothed but never deeply lobed. Upper leaves lack petioles, and sometimes their paired blade bases join,

Mohavea confertiflora (ghost flower)

surrounding the stem. The two-lipped corolla has a well-developed tube, with the staminode usually hugging the floor and the stamens often lining the roof, where anthers are positioned to deposit pollen on the backs of insect visitors. Sometimes the three lobes of the lower corolla lip project forward enough to provide a good landing platform for incoming pollinators. Corollas are frequently pink to purple or bluish, but the deserts also have several red species.

Penstemon floridus var. *austinii* (**Inyo beardtongue**) is spectacularly *floridus* (covered with flowers), growing about 1 m high and bearing numerous bright rose-pink corollas that are almost 3 cm long and glandular both inside and out. In this species the staminode is not hairy. Inyo beardtongue is found in northern desert mountains and the Inyo Mountains, at elevations of 1,000–2,400 m.

Penstemon palmeri (**scented penstemon**) is another tall (to 2 m) species with pink flowers, but these are light pink with darker lines. The floor of the corolla has some long white hairs, and the staminode is densely covered with yellow hairs that look like they might be attached to the corolla. Large black bees love

Penstemon floridus var. *austinii* (**Inyo beardtongue**)

this flower and are attracted by its fragrance. Occasionally you may be lucky enough to see a large patch of *Penstemon palmeri* in a roadside runoff area or in a wash of the eastern Mojave at elevations of 1,100 to 2,300 m. Dr. Edward Palmer (1831–1911) was an ethnographer and botanical explorer who collected thousands of plant specimens, mainly in the southwestern United States and Mexico. He has done well to have this spectacular *Penstemon* named for him.

Penstemon utahensis (**Utah firecracker**) is one of the red *Penstemon* species of our deserts, occurring only in the eastern mountains and points east, like

Penstemon palmeri (**scented penstemon**)

Penstemon utahensis (Utah firecracker)

Utah. Firecracker describes the scarlet red of the corolla and the long (about 2.5 cm) corolla tube. There are usually some white lines near the mouth of the tube, and lobes of both lips spread widely.

Scrophularia (**figwort**)—Although this is a large genus with about 250 species, only five occur in California, and only one in our deserts. The square cross section of stems gives away the genus. All of the species are perennials or shrubs covered with hairs and glands. Leaves are opposite, with regimental attention to adjacent pairs being oriented perpendicular to one another.

Scrophularia desertorum (**desert figwort**) is "of the desert" (*desertorum*), growing in the northern desert mountains and dry eastern Sierra Nevada at elevations of 1,000–3,000 m. Its corolla is about 1 cm long with an almost spherical tube and an upper lip that is maroon and flared back. The lower lip is tiny and, with the lower portion of the tube, light colored.

Veronica (**speedwell, brooklime**)— This genus named for Saint Veronica has about 250 members, mainly in Eurasia. Of the 16 species found in California, over half are nonnative, including one of the two species occurring in our deserts. *Veronica* species are annuals or perennials with a preference for wet places like streams and brooks. Desert species are uncommon and live only in wet mountainous areas. Leaves are opposite, and plants often spread by rooting at leafing points on stems that touch the ground. Flowers are unusual for having a four-lobed calyx and a four-lobed corolla with a very short tube. The upper corolla lobe is actually two fused lobes and forms the upper lip, while

Scrophularia desertorum (desert figwort)

Veronica americana (American brooklime)

the other three lobes form the lower lip of the two-lipped, bilaterally symmetrical corolla characteristic of scrophs. The two stamens that project forward of the corolla lobes above the middle of the flower are distinctive and will remind you that this flower has bilateral symmetry, even though the corolla lobes are almost radially symmetrical.

Veronica americana (**American brooklime**) is the patriotic native of California's desert mountains. Look for it up to 3,300 m, where it might blend into stream bank vegetation. The blue to slightly violet corollas are sometimes hard to spot.

Examples and Uses

At least one genus of Scrophulariaceae has medicinal uses other than shrinking hemorrhoids and chasing off lice.

Digitalis (foxglove) produces cardiac toxins that, used in extremely controlled, tiny amounts, slow heart rate, sometimes resulting in a stronger beat. Because the amount of poison that an herbivore such as a deer or rabbit would get from browsing on *Digitalis* would be fatal, you will find foxgloves standing tall after the lilies and buttercups have all been eaten off.

Species of at least 64 genera of scrophs are used as ornamental plants. Beardtongues (*Penstemon*) are particularly popular in western gardens where drought-tolerant species provide large floral displays year after year. Many showy varieties of foxglove do well in shady locations and reseed reliably, in addition to being resistant to garden grazers. Snapdragons (*Antirrhinum*) come in a huge variety of colors and brighten early-season gardens when temperatures are still cool. Your local nursery will have many other examples.

Physalis crassifolia (thick-leaved ground-cherry)

SOLANACEAE

NIGHTSHADE OR POTATO FAMILY

How can this toxic family supply us with so many familiar foods, such as tomatoes, eggplants, peppers, and potatoes? Tobacco is another well-known member of Solanaceae, and it makes sense that nicotine, or rather *Nicotiana*, is the genus name for tobacco. Native Americans were the first to discover many of the different uses for various plants in this family, and others have followed eagerly in their footsteps.

The Latin *solamen*, meaning "quieting," is the root for both the family name and the genus name *Solanum*. Quieting or narcotizing does describe the physiological effect of many members of this family. The toxicity of the family is perhaps referenced in the common name, nightshade, as in "being poisoned in the shade of the night" or "having a nightlike shade fall over you from poisoning."

Solanaceae Icon Features

- Radial symmetry with five petals joined to form a dish or tube, often with five points
- Five anthers, usually large, often in a tight ring
- One style and stigma with two lobes

Distinctive Features of Solanaceae

Solanaceae members have a tubular or dish-shaped corolla with five lobes. Occasionally, there are only four lobes, and in some genera, such as *Lycium* (thornbush), individual plants may have flowers of both types. At the bud stage, the corolla is often folded or pleated so that it opens like an upside-down umbrella with only five ribs. Sepals in Solanaceae are also joined to form at least a small tubular part with five (rarely four) lobes, reflecting the disposition of the petals.

The calyx in Solanaceae may be either diminutive or domineering. Some genera are characterized by a calyx that expands throughout maturation of the fruit, forming an outer layer like a slipcover or cloak—as in tomatillos (*Physalis philadelphica*). Though not characteristic of the entire family, an enlarged, enveloping calyx, when present, definitely suggests Solanaceae.

The five anthers characteristic of nightshades are generally large and often tightly arranged around the style, as if needed to hold it up straight. Sometimes anthers open just at the tips to let pollen out of pores or small slits. In other cases, they open by long slits from the tip nearly to the base. Filaments may be either longer or shorter than the anthers, but they are always attached to the base of the corolla tube between the lobes.

There is just one style in Solanaceae, but the stigma may be two-lobed, reflecting a two-chambered ovary. The horticultural tendency to keep enlarging fruits may make it difficult to believe that the original wild tomato or pepper had just those two ovary chambers. Check this on some native species, but do not eat them! The superior ovary is evident in garden tomatoes: the green calyx lobes at the stem end show that the ovary (later the fruit) was positioned above, or superior to, the attachment of other flower parts (of which only the calyx remains).

Solanaceae members come in virtually every form from annuals to trees and vines, but in the California deserts the only tree, tree tobacco (*Nicotiana glauca*), is a nonnative, and there are no vines. Leaves of Solanaceae are usually alternately arranged on the stem, and they have petioles but no stipules. Woody plants are frequently thorny, and herbaceous plants are often glandular. Some plants have a bad odor, similar to tobacco, and all should be considered poisonous, except for a few fruits or tubers, as noted.

Similar Families

There are many families with radial symmetry, five fused petals, five fused sepals, and five stamens with their filaments attached to the corolla tube. The umbrellalike pleating of the petals in Solanaceae, though a good indication of the family in general, is not evident in many of our desert species. Moreover, similar folding occurs in the Morning Glory Family (Convolvulaceae), though as a rule this family has a twining vine for a stem, a feature never found in California desert Solanaceae species. Concentrate on the female parts for the final distinction, along with some hints from stems and leaves.

Solanaceae always has a superior ovary—although in this it is not unique. Further, Solanaceae has only one style with no branching. This will rule out the Phlox Family (Polemoniaceae), with three style branches, and the Waterleaf Family (Hydrophyllaceae), with two style branches. It also rules out the one desert genus in the Morning Glory Family that does not have a twining stem.

Alternate leaves without stipules are an almost universal requirement for desert Solanaceae membership, allowing quick elimination of the Vervain Family (Verbenaceae), which has opposite leaves. Only one desert Solanaceae species, *Petunia parviflora*, could be mistaken for having opposite leaves, but they are not consistently so throughout the plant.

In some respects, including having a single style, flowers of the Borage Family (Boraginaceae) are similar to those of Solanaceae, but the possibility of confusing them is minimal. Almost all desert species of borage have flowers that are lined up along one side of a coiled tip of the stem: the shape of the inflorescence is that of a scorpion's tail. This is never the case with Solanaceae. In rare instances where the shape of the inflorescence does not make the distinction, it might be necessary to check the fruits. Borage Family fruits come apart in four separate segments attached only near the base. This is very different from the two solidly connected chambers of Solanaceae fruits.

Family Size and Distribution

Solanaceae comprises 75 genera and about 3,000 species distributed throughout the world, with some preference for the American tropics. The largest genus by far is *Solanum* (nightshade), with about half the total number of species, including potatoes (*Solanum tuberosum*), and eggplants (*Solanum melongena*).

California has 12 genera of Solanaceae, four of which are entirely nonnative, like the tomato (*Lycopersicon esculentum*), which readily escapes from cultivation and enters the wild. Seven of these 12 genera are found in California's deserts, with all of them having one or more native species. Although the larger genera include several desert nonnatives, in general the weedy species do better in other habitats.

Despite the family's preponderance of tropical species, Solanaceae members in California's deserts are well adapted to extremely arid conditions, requiring neither springs nor streams to survive. Many are widespread throughout our deserts, while others are found in harsh habitats of the southern deserts but do not tolerate frost. Only a few are confined to desert mountain habitats.

California Desert Genera and Species

California desert Solanaceae species come in almost every life form except vines and large trees. Similarly, flowers can be small (5 mm) to large (almost 20 cm), with a similar size range for leaves. White is a fairly common flower color, often tinged with purple or green, but several species have yellow flowers. Solanaceae fruits are frequently fairly small (1 cm or less) and round and smooth, but there are distinctive exceptions.

SIMPLIFIED KEY TO DESERT GENERA OF SOLANACEAE

Shape of the corolla and size of the calyx in fruit are important features for distinguishing among genera. Flower color and fruit features also are useful, as is the growth habit of the plant. It is ideal to find specimens with both fresh flowers and mature fruits.

1. Flowers saucer- to bowl-shaped. Corolla tube spreading from the base.
  2. Calyx small or bell- to bowl-shaped at fruit, not fully enclosing fruit (5 spp) .... *Solanum*
  2' Calyx enlarging with fruit and nearly surrounding it.
    • Corolla yellowish (6 spp) .... *Physalis*
    • Corolla mostly white .... *Chamaesaracha coronopus*
1' Flowers funnel- to trumpet-shaped. Corolla tube long and narrow at the base, lobes often spreading.
  3. Flowers large (more than 10 cm long). Fruit large (approx. 3 cm) with stiff prickles (2 spp) .....
  .... *Datura*
  3' Flowers medium to small (under 4 cm long). Fruit small (1 cm or less) and smooth or hairy, without prickles.
    4. Shrub or small tree.
      • Plants unarmed. Corolla yellow .... *Nicotiana glauca*
      • Plants usually armed. Corolla white to lavender, purple, or greenish (7 spp) .... *Lycium*
    4' Annual or perennial.
      • Plants low, spreading. Leaves mostly opposite, fleshy .... *Petunia parviflora*
      • Plants upright. Leaves alternate, not fleshy (4 spp) .... *Nicotiana*

Datura (**Jimson-weed, thorn-apple**)—This highly toxic genus, bearing an ancient Hindu name, has only about 13 species, but they occur throughout warm regions of the world, including India, where the name probably originated. Members of the genus were widely recognized by early civilizations for their toxic and hallucinogenic effects, which were sometimes incorporated into religious rituals. Ingestion of any part of any *Datura* species can be fatal; foolish people who are tempted to experience the hallucinogenic properties associated with this genus often wind up dead.

The common name thorn-apple describes fruits of the genus, which are fairly large and very prickly. In the late 1600s the common name Jimson-weed was applied to *Datura stramonium* because it was first observed near Jamestown, Virginia. The similar appearance of other North American species has led to widespread use of the name.

Datura, which may be annuals, perennials, or subshrubs, have rather large white to purplish funnel-shaped corollas that open mainly at night and are pollinated by large hawk-moths. In the California deserts, *Datura* often appear very lush, considering their surroundings. Foliage is grayish green, and large (for the desert) oval leaves appear in mounds to knee or waist height following the rainy season. The bad-smelling and toxic nature of herbage protects leaves from most herbivory (though various beetles love them), and fruits form before plants dry up from lack of soil moisture. Perennial species grow quickly from tubers following even brief rains and usually manage to flower even under the worst of conditions.

Datura discolor (small thorn-apple)

Datura discolor (small thorn-apple) is named for its "two" (*dis-*) distinct "colors" (*color*): white over the spreading part of the corolla, and deep purple in the narrow tubular part. It is an annual, smaller than *Datura wrightii*, and confined to the Sonoran Desert at elevations under 500 m.

Datura wrightii (western Jimson-weed) has spread to the deserts from more coastal locations and is now a common feature beside the road. It grows to over 1 m with large, dense, dark grayish-green leaves and huge (20 cm) white trumpet-shaped flowers that generally wilt during the day. Charles Wright (1811–1885) was one of the western plant collectors who sent specimens to Asa Gray at Harvard University.

Lycium (box-thorn, thornbush, wolfberry)—There are about 100 species of this woody and well-armed genus worldwide, occupying warm, dry climates. Nine species inhabit California, with seven of them in the deserts. The scientific name is derived from Lycia, a Classical Period country in the south of present-day Turkey, where species of the genus were well known. Two common names emphasize the rigid thorns, while the third probably reflects the fact that *lyc-* is the Greek root for "wolf."

Lycium species are shrubs with small (usually about 1 cm) linear to oval leaves. Funnel-shaped flowers, often with a narrow tube and flaring lobes, are also around 1 cm in length, and they may bloom profusely along stems that become leafy following rains. Flower colors include various shades of greenish and purplish white

Datura wrightii (western Jimson-weed)

to pink. Fruits are small (less than 1 cm) berries with two chambers. Most turn bright red at maturity, and unlike most other members of the family, *Lycium* produces fruits that are not poisonous. Mammals and birds feed on them extensively, and before grocery stores they were useful to Native Americans.

Lycium andersonii (**Anderson thornbush**) is the most widespread of the California species, with large expanses growing on many gravelly and rocky slopes under 1,900 m. It was named after Charles Lewis Anderson (1827–1910), a physician and botanist who is honored in the names of several common western plants. The leaves on his species of *Lycium*

Lycium andersonii (Anderson thornbush)

are 3–15 mm long and very narrow but fat, so that the cross section is a tiny (2–3 mm) oval. The corolla is about 1 cm long, mostly a narrow tube, with either four or five lobes of less than 1 mm each. The calyx is just a tiny cup at the base of the corolla tube. Flowers are usually white to violet, and fruits red to orange.

Lycium brevipes (**frutilla**) has an attractive deep pink to lilac flower with a wider tube and larger lobes than its cousin *Lycium andersonii*. The four or five lobes are 3–5 mm long with stamens extending well into view. Leaves are approximately oval, up to 15 mm long, and covered with tiny glands and microscopic hairs. *Brevipes* means "short-footed," referring to the flower pedicel, which is not, however, always short. The name frutilla, sometimes applied to the whole genus, probably refers to the fact that the berries can be considered "little" (*-illa* in Spanish) "fruits," in the sense of being small and edible. *Lycium brevipes* grows below 600 m in the western Sonoran Desert.

Lycium cooperi (**peach-thorn**) is named for Dr. J. G. Cooper, a geologist who collected plants in the Mojave Desert in 1861. Peach-thorn is a stout-stemmed thorny shrub with oval leaves to 3 cm and

Lycium brevipes var. *brevipes* (frutilla)

numerous greenish white flowers. It grows widely throughout the Mojave Desert under 2,000 m.

Lycium pallidum var. *oligospermum* (rabbit-thorn) is called "pallid" for its gray-green leaves. Flowers have broad lobes that are heavily purple-veined. Branches are stiff, spreading, and thorny. It grows under 1,200 m in the Mojave and northern Sonoran Deserts.

Lycium parishii (Parish's desert-thorn) was named for Samuel Bonsall Parish (1838–1928) and perhaps his wife, who were pioneer botanical collectors in southern California. The species is rare in the state, found only on the western margin of the Sonoran Desert in a few canyons. It has numerous thin branches with small leaves. Flowers are lavender, and stamens extend well beyond the corolla tube.

Lycium cooperi (peach-thorn)

Lycium pallidum var. *oligospermum* (rabbit-thorn)

Nicotiana (tobacco)—Jean Nicot (1530–1600) was a French ambassador to Portugal, where he obtained tobacco seeds from early voyagers to the Americas and presented them to the queen of France. This notorious genus has about 60 species, most of which originated in the Americas. *Nicotiana tabacum* is now cultivated in many parts of the world, with

Lycium parishii (Parish's desert-thorn)

Nicotiana obtusifolia (desert tobacco)

the United States a major producer. All members of this genus have an odor similar to tobacco, and several species, in addition to *Nicotiana tabacum*, have been used for narcotic purposes. Leaves in this genus are entire, corollas are tubular, and calyces enlarge in fruit but do not fully enclose the fruits.

Of the seven species occurring in California, only four are natives, the others having arrived from South America. All of our native species, plus one nonnative, tree tobacco, can be found in our deserts. The native species are upright, glandular annuals or perennials (one is woody only at the base), with white to greenish flowers that have a long corolla tube with small lobes. Tree tobacco is the only really woody species that occurs in California, and its flowers are narrow yellow cylinders about 3 cm long.

Nicotiana obtusifolia (**desert tobacco**) was used as a narcotic by Native Americans. It is found throughout both deserts, growing best in rocky canyons and washes under 1,600 m. Although very glandular and sticky—your hands will smell bad if you touch it—it is often a handsome perennial, growing to roughly half a meter in height and well endowed with "blunt" (*obtus-*) oval leaves (*folia*) as much as 10 cm long. The flower is dull white to greenish, and the long (15–26 mm) funnel-shaped corolla has spreading rounded lobes. Flowers are loosely clustered along the upper ends of the branches, above most of the leaves.

Physalis (**ground-cherry**)—The common name comes from cherry-sized fruits growing low to the ground. The "cherry" is surrounded by an enlarged papery calyx, which gives the genus its scientific name, meaning "bladder" in Greek. The genus includes annuals to perennials and subshrubs. The corolla, which is usually yellowish, spreads from the base, forming a saucer or bell shape with five obscure lobes. The leaves may be entire to lobed, hairy or smooth, glandular or not, and opposite (almost) or alternate, but the fruits are very distinctive. Some of the ripened fruits are edible—

like tomatillos (*Physalis philadelphica*), relatives similarly enshrouded in a papery calyx—but the rest of the plant, including unripe fruit, is toxic. About 85 species are found in the Americas, Australia, and Eurasia. California has nine species, only four of which are native. All of the natives are found in our deserts, along with two of the nonnatives.

Physalis crassifolia (**thick-leaved ground-cherry**) is the most common and widespread of native *Physalis* species. It is a heavily

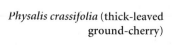

Physalis crassifolia (**thick-leaved ground-cherry**)

branched perennial, growing to almost 80 cm under favorable conditions, but usually about half that height. The leaf blade is usually a wide oval, 1–3 cm long, with a petiole that is as long as the blade. Leaves are covered with dense, short, unbranched hairs and tend to be somewhat fleshy (hence the specific epithet, meaning "thick-leaved") and glandular.

Solanum (**nightshade**)—This largely tropical genus is the largest of the family, with about 1,500 species. Many are toxic, but some provide important foods such as potatoes and eggplants—although even if some portion of the plant is edible, the remainder is likely to be toxic. The genus is represented in California's deserts by annuals, perennials, and subshrubs, but elsewhere there may be woody shrubs and vines as well. The flowers have a wide spreading corolla, usually with indistinct lobes like a five-sided plate. Sometimes the outer edges curve back away from the stamens, as if you were looking at a shallow bowl from the bottom. *Solanum* anthers are bigger than the filaments that support them, and they usually

Solanum douglasii (Douglas nightshade)

stand erect, edge-to-edge, in the center of the flower, with the style extending past them. Anthers open only by pores or short slits near the tip, where pollen is shaken out by wasps or flies. Fruits are usually spherical, about 1 cm in diameter, and green to very dark blue or purple. Of the 19 species occurring in California, only six are natives. Three of the natives and two nonnatives can be found in our deserts, although they are not generally very widespread.

Solanum douglasii (**Douglas nightshade**) was named for the Scottish botanist David Douglas (1798–1834), who collected in California in 1831–1832. Found under 1,000 m in the Mojave Desert in dry shrubland, it is a widely branching perennial or subshrub, usually about 1 m tall. The white corollas have long, somewhat pointed lobes, forming a star shape. The prominent anthers are bright yellow, and the berries green, turning almost black-blue.

Examples and Uses

There are dozens of varieties of potatoes (*Solanum tuberosum*), the ancestors of which were all originally native to Central and South America. Peruvian Indians cultivated some varieties 4,000 years ago, at elevations as high as 5,000 m, and they learned to freeze-dry them long before that method of preservation occurred to instant soup manufacturers. A potato is a tuber, the swollen end of an underground stem that acts as a storage organ for starch. It is the only part of the plant that is not toxic. However, even some parts of the tuber become toxic under conditions such as exposure to light and storage at very cold or overly warm temperatures, which promote the formation of alkaloids. The traditional potato cellar provided a dark, cool location, safe for long-term storage. Any green portion of a potato and any healthy-looking, sprouting "eyes" contain potentially dangerous levels of toxic alkaloids and should be thoroughly removed and discarded.

Potatoes, tomatoes (*Lycopersicon esculentum*), and peppers (*Capsicum annum* and *Capsicum frutescens*) all were originally cultivated in Central or South America and introduced into Europe in the late 1500s. Adopted first in England and Ireland, potatoes were found to be easy to grow and store in large quantities. The new source of food proved invaluable—until the potato famine of 1845–1849, when a serious blight wiped out crops. The rest of Europe was more skeptical of potatoes, based on the toxic reputation of the family, and adopted their use more slowly, never reaching the same level of dependence.

Capsicum peppers are now grown throughout the world in almost infinite variety and are used extensively in India and Southeast Asia as well as other regions where they are not native. Although peppers are not considered toxic, their heat is caused by an alkaloid called capsaicin, which causes burns and irritation in proportion to its concentration—a powerful deterrent to animals eating the fruit and a useful component of some very effective pesticides. Nevertheless, peppers are a popular ingredient in both condiments and main courses, and heat tolerance is sometimes a status symbol for those who eat them.

Tobaccos of various sorts have been made from many members of the genus *Nicotiana*, and tobacco was smoked by Native Americans throughout the continent, often just at ceremonial events. Desert tobacco (*Nicotiana obtusifolia*) was cultivated by the Yuma and Havasupai Indians, who scattered its seeds among the ashes of mesquite trees they had cut and burned. Indians of eastern North America had large-leaved species, such as *Nicotiana tabacum*, for smoking. European explorers were quick to sample and imitate the habit, and tobacco spread rapidly around the world, becoming a huge commercial crop. For decades the U.S. government supported research to improve varieties of tobacco and increase its productivity. Much of our understanding of plant physiology stemmed from this work.

Some of the same compounds that are toxic in various species of Solanaceae have been refined and used for medicinal purposes. Atropine, the drops that dilate your pupils for the eye doctor, is made from nightshade (*Atropa belladonna*). Stramonium, a historical treatment for sores and wounds, was prepared from the leaves or buds of *Datura* species and was probably learned from Native Americans. Some Native Americans used *Datura* ceremonially to induce hallucinations, and occasionally people still experiment for this purpose without realizing how easy it is to get a lethal dose. Even the hawk-moths that come to *Datura* for nectar and pollinate the flowers get intoxicated. However, when the moth's larvae feed on *Datura*, they sequester the alkaloids and become poisonous to their predators.

Finally, several members of the Nightshade Family are used chiefly as ornamental plants. The best known is probably the garden petunia (*Petunia*), which comes in many shades from white to purple and has a flower 6–8 cm across. Perhaps the most impressive landscape plant in the family is angel's trumpet, a small tree in the genus *Brugmansia*, that has long (20–30 cm), pendulous white to apricot flowers shaped like a trumpet. These are beautiful but deadly if eaten.

Larrea tridentata (creosote bush), Death Valley National Park, Mosaic Canyon

ZYGOPHYLLACEAE

CALTROP FAMILY

The relatively small Caltrop Family includes one of the most widespread, common, and drought-tolerant plants of California's deserts. Indeed, the Sonoran and Mojave Deserts together can almost be defined by creosote bush (*Larrea tridentata*). It grows over vast expanses, often attracting attention because of the relatively uniform distance between individual shrubs and because of its dominance in areas where almost nothing else grows. Creosote bush is extremely well adapted to the desert and has incredible staying power: some large circles of plants have been shown to be genetically identical offspring, or clones, from the germination and growth of a single seed thousands of years ago.

In Greek, *zygo-* means "joined" or "yoked," as in pairs of yoked oxen, while *phyll-* relates to "leaves"; so Zygophyllaceae means the family with yoked or paired leaves. Caltrop is a variant of a medieval Latin name for a spiny thistle, as well as for a weapon, and was applied to the Mediterranean genus *Tribulus*, the most infamous member of the family, endowed with serious spines.

Zygophyllaceae Icon Features

- Leaves once compound, arranged opposite
- Leaflets two (fused at the base or not), three, or 6–18
- Stems swollen at nodes (sometimes with spiny stipules at the leaf bases)

Distinctive Features of Zygophyllaceae

The joined or yoked pairs of leaves in Zygophyllaceae are once-compound leaves in opposite formation (paired). There may be two or three or many leaflets per leaf, usually in even numbers when many. Whatever the number of leaflets, they are all smooth edged, without teeth or lobes, and they are usually small (about 1 cm or less). Pairs of leaves are often unequal, with one being larger than the other. Sometimes leaves have tough, sharp stipules flanking the petiole's attachment to the stem. This results in four nasty spines at every leaf pair. Stems may be woody or not, but they are usually branched repeatedly, often in a zigzag pattern with swellings at the branching points.

Flowers in Zygophyllaceae have radial symmetry (sometimes a bit off), with flower parts generally in fives, though a few genera are a little sloppy with the numbers. The five sepals are usually separate from one another, as are the five petals. Petals are often clawed, or bent into an L-shape near the middle, and the spreading tips frequently twist like a pinwheel, or just like a bad hair day. There are 10 stamens, often with glands or appendages at the base of filaments. The ovary is superior and has 5–10 chambers, each producing one to several seeds. There is just one style, which may be ridged or angular, and the stigma may be lobed or rounded.

Fruits in Zygophyllaceae genera may appear very different, and they are usually quite distinctive—some with dangerous spines, others with cottony hairs. What fruits have in common is five (or occasionally 10) compartments that either split apart when ripe into five (or 10) separate pieces, or that look, on the outside, like lobes or ridges on the intact fruit.

Similar Families

The flower-part counts for Zygophyllaceae are similar to some flowers in the Rose Family (Rosaceae), but there is an obvious difference in the leaves. Rose Family members almost always have alternately arranged leaves rather than the opposite positioning of leaves in Zygophyllaceae. There is only one exception to this in desert roses: blackbush (*Coleogyne ramosissima*), which is also an exception to the flower-part counts, with no petals and four sepals, so you will not be looking at it and thinking caltrop.

The presence of oppositely arranged once-compound leaves greatly helps to separate Zygophyllaceae from other families that have desert members with five sepals, five petals, 10 stamens, and a superior ovary. This is true even for the Bean or Pea Family (Fabaceae), a few members of which (*Cercidium, Senna*) have flowers and compound leaves somewhat resembling a few genera of Zygophyllaceae. More important, the fruits of Fabaceae, even when immature, are easy to spot as beans (legumes), which are nothing like the rounded, five-lobed or compartmented fruits of Zygophyllaceae.

Family Size and Distribution

There are 26 genera of Zygophyllaceae with about 250 species. Most of them favor warm, arid regions of the world, and many of them (although not California's) are well adapted for saline habitats. *Zygophyllum* (bean-caper) is the largest genus, with about 80 species native to North Africa and Asia. Its only species in California (*Zygophyllum fabago*) is from the Mediterranean region and central Asia.

Of the five genera represented in California, two are entirely nonnative. The native genera all are small, having only one or two representatives in the state, all four of which can be found in our deserts (although one is rare). Creosote bush is one of the most obvious and common of all desert plants and is found at low elevations (under 1,000 m) throughout the Mojave and Sonoran Deserts. It arrived from South America about 15,000 years ago under mysterious circumstances, since there neither is nor was any suitable habitat for thousands of miles between the populations. Other species native to California generally are restricted to limited southern habitats and are hard to find.

California Desert Genera and Species

Zygophyllaceae members in California may be annuals, perennials, or shrubs, sometimes with spines at leafing or branching points or on fruits. Leaves all are compound and opposite, but differ in having two, three, or many leaflets. Flowers are yellow, orange, or pink to purple, and usually big enough to notice at a short distance. Some of the spiny fruits are especially noticeable in your foot, air mattress, or bicycle tire.

SIMPLIFIED KEY TO DESERT GENERA OF ZYGOPHYLLACEAE

Zygophyllaceae genera in California are easily distinguished based on characteristics of their leaves and fruits. If these features are absent, flowers or just stems may be sufficient for identification.
1. Leaves with 6–18 leaflets, stipules not spiny.
   • Fruit lobed, not spiny, breaking into 10 nutlets. Petals mostly orange (3 spp) .... *Kallstroemia*

- Fruit with stout spines, breaking into 5 nutlets. Petals yellow .... *Tribulus terrestris*
1' Leaves with either 2 or 3 leaflets, stipules spiny or not.
   2. Leaves with 3 spine-tipped leaflets, stipules also spiny. Petals pink to purple (2 spp) .... *Fagonia*
   2' Leaves with 2 leaflets, not spiny.
     - Shrub. Leaflets fused at the base, surface resinous. Petals yellow .... *Larrea tridentata*
     - Perennial. Leaflets not fused, fleshy. Petals yellow to orange .... *Zygophyllum fabago*

Fagonia—G. C. Fagon was a 17th-century French botanist and physician to Louis XIV. The genus named for him has about 18 species spread through several Mediterranean climate regions of the world. California has two species, both native and both living in our southern deserts. *Fagonia* may be either a shrub or a perennial, and its leaves have three leaflets with spiny tips and spiny stipules. Flowers are pink to purple and grow singly from the branching points or leaf axils. Fruits are five-lobed but not spiny.

Fagonia laevis (**smooth-stemmed fago-nia**) is a small, intricately branched shrub, with thin angular stems that are yellow-

Fagonia laevis (**smooth-stemmed fagonia**)

green and just barely seem woody. It is usually well under 1 m tall, and it grows on hot rocky hillsides and washes below 700 m in the Sonoran and southern Mojave Deserts. Flowers are about 1 cm across, with spreading dark pink to magenta petals. Leaves may be sparse, but even when present, the three leaflets are narrow (1–4 mm wide) and shorter (3–9 mm) than the petiole, so they do not look significant compared to the dense branches. Each leaf has two spiny stipules at its base. The Latin word *laevis*

means "smooth" and refers to the lack of glands or hairs on mature stems and leaves. This smooth quality helps distinguish *Fagonia laevis* from the other California species, *Fagonia pachyacantha*, which has obvious glands on older leaves and stems.

Fagonia pachyacantha (**spiny-stemmed fagonia**) is named for having a "thick" (*pachy-*)

Fagonia pachyacantha (spiny-stemmed fagonia)

"thorn" (*acantha*). Stems and leaves are thick also, in contrast to *Fagonia laevis*. *Fagonia pachyacanthus* is a perennial growing only under 500 m in the Sonoran Desert.

Kallstroemia—Johannes Antonius Scopoli (1723–1788), who originally authored this genus, gave it the name of his contemporary, Anders Kallstroem. There are 17 species, all in warm climates of the Americas, with one native and two nonnatives found in California. The native species, *Kallstroemia californica*, has not been collected from a wild population in the state since 1948, but it grows east to Texas and is in Mexico. *Kallstroemia* is an annual that spreads across the ground and has paired leaves with 6–18 leaflets. Stipules are long and thin but not spiny. Fruits are lumpy or warty and have 10 lobes that split apart at maturity. The two nonnative species have orange flowers, and *Kallstroemia californica* has yellow flowers.

Kallstroemia parviflora (small-flowered kallstroemia)

Kallstroemia parviflora (small-flowered kallstroemia) is native to Arizona and Mexico but not to California, where it is generally restricted to sandy slopes and roadsides. It has long sprawling stems with few leaves or flowers. The lumpy fruits retain the flower's style, which is longer than the fattened fruit body and persists even after the fruit segments have split away. *Kallstroemia parviflora* is "small" (*parvi-*) "flowered" (*flora*) with orange petals 6–12 mm long. This makes it easy to distinguish from its Arizona cousin, *Kallstroemia grandiflora*, which is "large" (*grandi-*) flowered with petals 15–30 mm long and orange grading to lighter shades near the tips. *Kallstroemia grandiflora*, sometimes called summer poppy, and *Kallstroemia parviflora* are uncommon in California, but grow widely to the east and south where there are more reliable summer rains.

Larrea (**creosote bush**)—There are just five species of *Larrea*, all in South America, with one of them represented in California. All *Larrea* are shrubs without spines. Their distinguishing feature is the odd leaf with just two leaflets fused together at the base like Siamese twins joined near the petiole. Petals are yellow, and stamens have toothed appendages near the base of the filaments. Fruits are five-lobed and hairy, splitting into five hairy segments at maturity. The genus was named for Juan Antonio de Larrea, a Spanish clergyman and friend of science. He would be proud of the amount of science that has gone into elucidating adaptations of our California species, *Larrea tridentata*, to its arid habitat.

Larrea tridentata (**creosote bush, greasewood**) seems poorly named. Creosote generally refers to a wood preservative, made from petroleum, that accounts for the bad odor of old telephone poles and railroad ties. It has nothing to do with our desert genus, except that some people with ailing olfactory organs apparently thought *Larrea tridentata* smelled like the wood preservative. Most people think it has a rather pleasant, spicy odor, contributing to the invigorating scent of rain in extremely dry habitats. Greasewood is the common name usually applied to *Sarcobatus vermiculatus*, a salt-tolerant shrub with fleshy leaves, in the Goosefoot or Saltbush Family (Chenopodiaceae), and not a very good name for *Larrea tridentata* either. Finally, the specific epithet, *tri-* plus *-dentata*, says this is the "three-toothed" *Larrea*, which is hard to see since it refers to irregularly toothed appendages of the filament bases.

In spite of any hardships related to naming, *Larrea tridentata* is very common and successful as

Larrea tridentata (creosote bush), flowers and cottony fruits

a desert species. It outstrips cacti in its drought tolerance and has a distribution that almost defines the Sonoran and Mojave Deserts. With the exception of elevations over around 1,000 m and highly saline flats, it occurs virtually everywhere, even in places where nothing else will grow. It can live in the hottest valleys for two to three years without any rainfall, though it looks terrible; the leaves, which are normally a shiny varnished-looking yellow-green, turn rust brown to almost black and then fall off, sometimes along with small twigs. Shortly after a good rain, however, plants completely restore their leaves and burst into bloom with 1 cm yellow flowers. Flowers are followed by cotton balls the size of a large pea. These are formed by dense white to silvery hairs growing on the fruits.

The size and spacing of *Larrea tridentata* shrubs is a fair indicator of rainfall. In very dry places, height is restricted to well under 1 m, and plants are widely spaced. Where there is more soil moisture, plants may grow to 2 m and be more closely spaced. Creosote bush roots secrete an inhibitor to germination of seeds of their own species. This prevents young plants from becoming established within the broad circular root zone, accounting for the even spacing between plants. This phenomenon does not interfere with rejuvenation of the population because new stem growth arises from the outer edges of a plant's root crown; the plant thus expands into a circle with the center eventually dying out. Over the centuries, or even millennia, as the circle of plants widens, individuals become separated, but they are all genetically identical to the original plant. If these were animals, we would call it cloning, but in plants it is just vegetative reproduction, a common practice for especially successful plants that do not need to bother with exchanging their genetic material through sexual reproduction.

Another interesting feature of *Larrea tridentata* is its close relationship to *Larrea divaricata*, its Argentinean sister, from whom it differs only by having double or triple the normal number of chromosomes. The more heavily endowed plants (with triple the normal) grow farther north in

North American populations than do the populations with only double the normal. The explosive success of the North American species, over only 15,000 years, may be related to greater adaptability somehow conferred by the extra—but identical—genetic material.

Tribulus (**puncture vine, caltrop**)—As the common name puncture vine suggests, this is a genus with piercing spines on the fruits. The scientific name, *tri-* (three) *bulus* (bulbs), alludes to these also—specifically, to the three-lobed spikes (sometimes two- or four-lobed) on each of the five fruit segments. Caltrop is the name for a spiked ball placed on the ground to puncture enemy tires or injure charging horses. Caltrops were invented in the Middle Ages and were named for the plant with similar (but smaller) fruits. However, the plant itself was probably named for its slim similarity to a spiny thistle called *calcatrippa* in Latin. In any case, the spikes, looking something like rhinoceros horns, are very effective at penetrating almost anything, and so have helped to spread fruits around the world. There are 12 species, most of which are native to North Africa and southern Europe, but several of which are now widespread as aggressive weeds.

Tribulus terrestris (puncture vine), flower above spiny fruit

Tribulus terrestris (**puncture vine, goat head, caltrop**) appropriately carries the common names referring to spiny fruits of the genus. In addition, it is toxic to livestock that eat the leaves. Following its arrival, it aggressively invaded dry habitats throughout large portions of the United States and Mexico, where it was considered a pernicious weed. It has now been controlled by introduction of a weevil that feeds on its seeds. The specific epithet, *terrestris*, means "growing on the ground or soil" (rather than on rock, for example), and may refer to the growth habit of spreading prostrate across the ground. Flowers are yellow and rather pretty, but small (less than 1 cm across).

Examples and Uses

Numerous uses have been found for the ubiquitous creosote bush. Before drugstores, people had a long list of medicinal uses, from wound dressings to cures for tuberculosis. Although these applications may have been perfected by Native Americans, based on centuries of tradition, none has proven benefits that would outweigh the hazards of self-administration. Sale of creosote bush leaves for tea has been banned due to several deaths, but some people still make their own concoctions. A better use of it is as an antioxidant added to paints, or in pharmacological research.

In most places, creosote bush is the tallest thing around, and its sparse shadow is enough to serve as a protective nursery for selected other species. These include cacti, whose root systems are shallow and do not compete significantly with the much deeper and more widely spreading roots of the creosote bush. Also, annual plants, sprouting only after significant rainfall, often grow especially well in the shelter of creosote bushes. This growth and decay cycle contributes to the accumulation of organic litter, some of it stabilized from wind distribution by multiple creosote bush stems, making a slightly richer soil under the shrubs than out in the open.

Naturally, many insects utilize creosote bush and pollinate its flowers, and some depend on it exclusively. Few mammals or birds feed on it, though desperate jackrabbits have been seen eating its

Larrea tridentata (creosote bush) with *Phacelia distans* laced throughout its protective cover

leaves, and small rodents like to burrow near the root crowns, perhaps benefiting from structural support for their tunnels in the sandy soils.

Other desert species of Zygophyllaceae are not very popular except for providing color in dry gardens. At least two South American genera, *Bulnesia* and *Guaiacum,* are prized for their very hard wood. The latter is also known as lignum vitae, meaning "wood of life," because its resin is used medicinally. It is also grown commercially for timber.

The authors with *Opuntia basilaris* var. *basilaris* (beavertail cactus) in Anza Borrego Desert State Park.

PHOTOGRAPH BY CAMILLE MORHARDT

ABOUT THE PHOTOGRAPHS

The photographs in this book were all taken on 35 mm film in the field using natural light. We used Kodak E200 Ektachrome—sometimes pushed to ISO 600 when the wind was strong and the flowers wouldn't stand still—Fujichrome Provia 100 F when things were calmer, and Fujichrome Velvia when calm enough for an ISO 50 film. We used several Canon lenses and bodies, but most commonly a 100 mm macro lens mounted on an EOS 1N or 1V body. The next most-used lenses were a 50 mm macro and a 45 mm tilt-shift. Most of the time the camera was mounted on a tripod, but sometimes lying on the ground and becoming a human tripod worked best in spite of the ubiquity of thorns, ants, and other dangers that frequent the desert floor.

Oenothera deltoides (dune evening primrose), Anza Borrego Desert State Park

ANNOTATED REFERENCES

This section lists the books we used frequently as references during the preparation of our manuscript. We describe the general sort of information most useful to us, and some of the best features of the work, in our opinion.

Baldwin, Bruce B. et al., editors *The Jepson Desert Manual* University of California Press, Berkeley, California, 2002. Copyright 2002, by The Regents of The University of California. The manageable size and thorough coverage of this work covering southeastern California make it a pleasure to use in the desert. It is essentially a subset of plants included in The Jepson Manual (Hickman 1996), using updated information on distribution to address only those species found in the Mojave and Sonoran deserts and in southern portions of the Great Basin Floristic Province. The format and most of the text are identical to the larger work. Over 100 color photographs have been added in a stand-alone central section.

Borror, Donald J. *Dictionary of Word Roots and Combining Forms* Mayfield Publishing Company, Mountain View, California. Renewed copyright 1988, by Arthur C. Borror. This pocket-sized listing of Greek and Latin words gives only the most brief translation possible, but it covers a lot of material for its size and convenience. Root words from many other languages also are included, and there is a good section on pronunciation.

Bowers, Janice Emily *Flowers and Shrubs of the Mojave Desert* Southwest Parks and Monuments Association, Tucson, Arizona. Copyright 1999, by the author. Text and line drawings (by Brian Wignall) in this small paperback all are black, and the pages are colored to correspond to flower colors. About 130 diverse desert species are covered, one to a page, with interesting text.

Dodge, Natt N. *Flowers of the Southwest Deserts* Southwest Parks and Monuments Association, Tucson, Arizona, copyright 1985. This is another in the series of black ink on colored papers corresponding to flower color, although there are a few color photographs of species in separate sections. Drawings are by Jeanne R. Janish. Many of the species included do not grow in California, but two or three related species frequently are treated on a single page describing differing distributions.

Hickman, James C., editor *The Jepson Manual* University of California Press, Berkeley, California, 1996. Copyright 1993, by The Regents of the University of California. The dog-eared appearance of our copy of this 1400 page book speaks for its usefulness in the field, classroom, and study. We have taken it as the authoritative source of information about plant names, distributions, dimensions, and floristic features, and we hungrily awaited its new companion, The Jepson Desert Manual (Baldwin et al. 2002). Every plant native to, or independently established (proliferating outside of cultivation) in California is included in this tome, so text about each species is very brief, and introductions to families are similarly cryptic. Small line drawings are provided for many species, and technical language is restrained but still makes reading appeal mainly to technical users.

Jaeger, Edmund C. *Desert Wild Flowers* Stanford University Press, Stanford, California, 1940. Copyright 1940, by The Board of Trustees of Leland Stanford Junior University. For many years this has been the most comprehensive and accessible work restricted to California desert plants. It treats 764 species with line drawings of most, arranged by family, with families in evolutionary order (now understood differently). Every species is listed with a common name, real or imagined. Descriptive text is very brief but often includes interpretation of Latin names, Native American uses of plants, insect relationships with plants, and plant distribution as understood at the time. There is no color, and few photographs. A key to genera has been added to later editions. There have been numerous name changes since the book was published but most are readily resolved, and the book is still extremely useful.

McGee, Harold *On Food and Cooking* Charles Scribner's Sons, New York. Copyright 1984, by the author. Fascinating information about all sorts of foods crams nearly 700 pages. There are excellent descriptions of fruits, vegetables, grains, and spices, explaining origins in cultivation, anatomy, culinary uses, nutritional value, processing requirements and many other interesting facts. Botanists and cooks alike will learn a lot.

McMinn, Howard E. *An Illustrated Manual of California Shrubs* University of California Press, Berkeley, California, 1974. Copyright 1939, by the author. Many California desert flowering plants are shrubs, and this book has good descriptive information, including many line drawings. Organization is by family, with keys to genera and species. There is also an overall key to genera as well as to families. The introduction provides good explanations of descriptive terminology.

Munz, Philip A. *A Manual of Southern California Botany* Claremont Colleges, Claremont, California, copyright 1935. Even though there are much more recent and comprehensive treatments by Munz, this remains our favorite due to its age and significance for the Claremont Colleges. Line drawings are lovely and conveniently dispersed among text rather than relegated to collections on a single page. Beware of name changes.

Munz, Philip A. *California Desert Wildflowers* University of California Press, Berkeley, California, 1962. Copyright 1962, by The Regents of The University of California. This book was a welcome companion to botany majors at Pomona College when we were students there—a refreshing addition to our technical tomes. The original cover says "96 color photographs, 172 line drawings, 2 maps, $2.95", and worth every penny. However, species are arranged by flower color, with all photographs in a separate section, and there is no discussion of family or genus features, nor are there any keys. Nevertheless, it is a handy little book that still is being reprinted. An updated version is on the way.

Niklas, Karl J. *The Evolutionary Biology of Plants* University of Chicago Press, Chicago, Illinois. Copyright 1997, by The University of Chicago. This fascinating book gives the broadest possible perspective on adaptive evolution of plants from the very beginning. A few examples, familiar to desert-lovers, are discussed.

Nilsson, Karen B. *A Wild Flower by any other Name* Yosemite Association, Yosemite National Park. Copyright 1994, by Nils Nilsson. This short book contains a collection of biographical essays about many early explorers of western North America after whom plants have been named. Sketches of plants and photographs or portraits of most subjects are included.

Phillips, Steven J. and Patricia Wentworth Comus, editors *A Natural History of the Sonoran Desert* Arizona-Sonora Desert Museum Press, Tucson, Arizona and University of California Press, Berkeley, California. Copyright 2000, by the Arizona-Sonora Desert Museum. With over 600 pages, this book covers a lot of natural history, including descriptions with line drawings of representative plants from 24 families. A few color photographs are grouped into a single section.

Raven, Peter H. and Daniel I. Axelrod *Origin and Relationships of the California Flora* University of California Press, Berkeley, California, 1995. Copyright 1978, by The Regents of The University of California. This technical work describes evidence and reasons for the richness of California's flora, including the large numbers of species that have evolved here and grow nowhere else.

Stearn, William T. *Botanical Latin* Timber Press, Portland, Oregon, 1998. Copyright 1992, by William T. Stearn and Eldwyth Ruth Stearn. This book is targeted to botanists who are involved in naming plants, and it goes into detail about how to prepare Latin descriptions, how to Latinize place names, and the like. There is a vocabulary section emphasizing how root words have been used, and there are chapters giving good descriptions of the subtleties of words describing color, shape, etc.

Stearn, William T. *Stearn's Dictionary of Plant Names for Gardeners* Cassell Publishers Limited, London, 1996. Copyright 1996, by W. T. Stearn and E. R. Stearn. The heart of this book is an alphabetical listing of genus names and specific epithets with their root words translated into English. Occasionally, Stearn includes a reason why the terms were selected and applied to particular plants. The introduction includes a good explanation of how plant names are formed and the rules that prevail.

Stewart, Jon Mark *Colorado Desert Wildflowers* Palm Desert, California. Copyright 1993, by the author. Beautiful photographs are the main reason to own this book. Species are arranged by color, with a full page, mostly occupied by a photo, for each. A good cross-section of the most obvious desert flowers, plus a few obscure ones are included, but there is very little botanical information.

Stewart, Jon Mark *Mojave Desert Wildflowers* Albuquerque, New Mexico. Copyright 1998, by the author. Photography is the highlight of this book, although it is not so impressive as in *Colorado Desert Wildflowers*, because of the smaller format. Coverage extends beyond California and is necessarily superficial. Arrangement is by flower color and text is just a few sentences per species.

Taylor, Ronald J. *Desert Wildflowers of North America* Mountain Press Publishing Company, Missoula, Montana, 1998. Copyright 1998, by the author. The broad coverage of this book is interesting but limits its usefulness for California. There are color photographs throughout, with text that emphasizes distinguishing features of species, and it is arranged by plant families. The back pages are devoted to a long key to families and some diagrams of plant parts.

Walters, Dirk R. and David J. Keil *Vascular Plant Taxonomy* Kendall/Hunt Publishing Company, Dubuque, Iowa, copyright 1996. This textbook stresses terminology in its coverage of vascular plant families, and includes good line drawings.

Woodland, Dennis W. *Contemporary Plant Systematics* Andrews University Press, Berrien Springs, Michigan, copyright 1997. This is an affordable and focused introductory text that includes brief descriptions of key features of all major groups of vascular plants. A full page is devoted to each of around 240 plant families, and there is good introductory material in initial chapters describing important anatomy of major groups. Final chapters of the book are interesting summaries of related topics. Illustrations all are black and white but there is a CD inside the back cover.

Zomlefer, Wendy B. *Guide to Flowering Plant Families* University of North Carolina Press, Chapel Hill, North Carolina. Copyright 1994, by the author. Stunningly beautiful and accurate drawings covering 130 plant families are the highlight of this book. You will not regret the absence of color. Text is also beautifully written and informative, although it demands an excellent command of technical terminology. Our favorite sections are those describing pollination mechanisms. There is a long, illustrated glossary, worth the price of the book all by itself.

Opuntia ramosissima (pencil cactus), Mojave Desert east of Kelso Dunes

INDEX